FRENCH PILOT

Volume Four
Le Morbihan to La Gironde

MALCOLM ROBSON

Nautical Books
Macmillan London

ISBN 0 333 32926 0

First published in Great Britain 1983 by
NAUTICAL BOOKS
an imprint of Macmillan London Ltd
4 Little Essex Street
London WC2R 3LF

Associated companies throughout the world

Caution
While great care has been taken in the compilation
of this book, neither the author nor the publisher
can accept responsibility for any inaccuracies and
omissions or mishaps arising from the work.

Photoset by Rowland Phototypesetting Ltd
Bury St Edmunds, Suffolk
Printed in Hong Kong

Contents

3

Contents – Continued

Introduction

Facing the prevailing SW winds, Biscay, land of fierce seas and strange legend, smiles under a reputation many nautical miles from the truth. Seen those sheltered rivers? Been up those sandy estuaries? Ever tried those busy little fishing ports? Sailing from the Channel, once round the corner of Ouessant and through the Raz, suddenly all is different . . . the weather pattern changes, the sun is hotter, seas are bluer.

This guide is for the chap who wishes to wander along the coast, among the islands, up the rivers; leisurely cruising. It is not for the owner who uses his boat merely for transport, or goes out for a day's sail to return to his home harbour at nightfall. Nor is it for those more modern sailormen whose seafaring is chained to the soullessness of a marina. So why not buy a few French charts, fit a pair of legs to your boat and get a close-up of those entrancing places you have been seeing only through binoculars? Within a week or so you will be speaking French and—if merely in self-defence—enjoying garlic. 'Every civilised man', reminds Ben Franklin, 'has two homelands. And one of them is France.'

The French treat electric supply as a jest, and sewage not at all, but when it comes to marking their coasts I know of no other country to equal theirs. The lights are numerous, the buoys enormous, the beacons prolific and anything with enough space carries its name in big letters. First class maintenance of the entire system is the year-round job of the *Ponts et Chaussées baliseurs* whose skippers incidentally are mines of information.

There has been a dramatic change in the atmosphere along the French coast toward *plaisanciers*, the direct result of the atomic proportions of the boating explosion. The marinas are filled as fast as they can be built and as a result they are wastelands of parked vessels often unused, seldom provide places for visitors and are somewhat expensive. Fishing harbours which formerly welcomed, or at least tolerated, visiting boats now sternly segregate amateurs from professionals. There remain all those small ports, many drying, some difficult to enter, but all uniformly glad to see the occasional visiting yacht.

You won't find this a treatise on navigation; locating your position before looking for pilot marks is elementary coastal plotting. Also, since inflation is as sure as sunrise, costs haven't been mentioned. Nor does this pretend to be a restaurant guide: what was top last year may be a *discothèque* this. But if there is one thing I would most humbly beg you to accept advice about, it is . . . legs. Every fishing boat, every yacht, motor boat, from 1 to 100 tons fits *les béquilles*. Fewer than one in ten of the harbours I describe doesn't dry, so without legs and a straight keel for taking the bottom you might just as well save the cost of this book. Since they were built before divers, few older harbour walls project further than LAT, so you will just have to get used to berthing single-legged, either against a quay or another boat. In any port, naturally, working fishing boats come first, so you will often have to dry in mid-harbour using both legs. You will therefore need a ladder: either folding, made from rope, or rigid. My own home-made one is typical of thousands: it fits to a leg when dried out in the open; it hangs from a ladderless (therefore often vacant) part of a high quay; it is used for bathing; it is a gangplank.

Notice how sparing I am about amenities. Whether the banks open on Friday or Saturday seems of less importance than knowing there will be 4·1 metres in an hour alongside the quay.

6

Lobster pots need particular care since they have floats on long buoyant rope, to cope with the tidal range. Often their owners lay them in narrow, rocky channels and with scant heed to traffic. And from traffic so to transport. After years of cruising with collapsible bikes we now have folding mopeds. With either you will see something of rural France apart from quaysides, thus making those remote anchorages even more pleasant. What's worse than a two-mile heavily laden walk in the opposite direction to a café?

Customs now oblige pleasure boats to use fuel taxed at the same rate as road users, and as quayside pumps only provide low-duty fuel to fishermen, this can become a problem. If you can take enough you might persuade a small road tanker to deliver by hose, otherwise take your cans to the filling station. Fuel Oil *Domestique* is forbidden for yachts.

Unless you use the tiny but ubiquitous International Camping Gaz cylinders, bottled gas is another perennial headache. Nobody will exchange British containers and I know of no handy plant where they can be filled, so the best solution is to carry enough spare gas on board. Paraffin users will laugh coarsely at all this fuss—until they too have to gather a few sticks for cooking—their tipple is called *pétrole* and may be bought with increasing bother at paint shops.

The information has been arranged from the north-west, Le Morbihan to La Gironde, and for convenience the area is chopped into three parts, see Fig. 2. There is no overlap either in charts or in marks.

French and Breton words in the text, unless they are names, are printed in italics, with a translation in the Glossary if required.

Finally, while information has been checked, sorted and rechecked, still errors can creep in so neither the publisher nor I can be responsible for mistakes or omissions. But, please, if you find any faults or you can supply additions, could you write to me in Sark?

Charts

Any mention of charts in this book refers to my sketch charts. The three areas J–L are in Fig. 2: detailed charts are indexed in Figs. 3, 4, 5. Charts are metric and soundings, drying heights and elevations have been reduced to LAT. Liberties have been taken to envelop collective dangers and there are only three basic contours: HAT, LAT and the 3-metre line. Anything below this can't be of great interest to yachts. In the text a figure in italics and underlined (e.g. *1·2m*) shows the height in metres drying out above chart datum (LAT). One

Fig. 1. Key to chart symbols

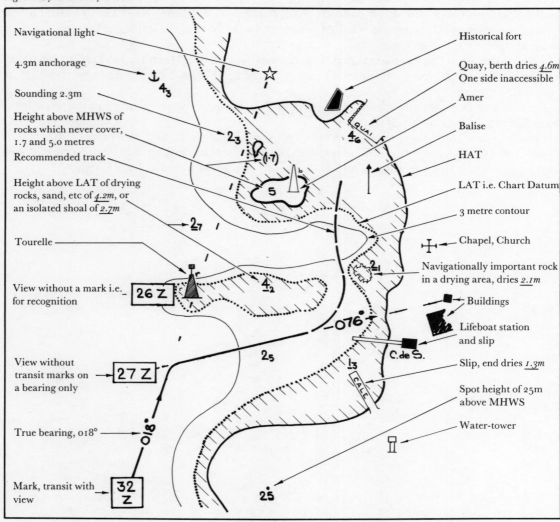

Navigational light

4.3m anchorage

Sounding 2.3m

Height above MHWS of rocks which never cover, 1.7 and 5.0 metres

Recommended track

Height above LAT of drying rocks, sand, etc of *4.2m*, or an isolated shoal of *2.7m*

Tourelle

View without a mark i.e. for recognition

View without transit marks on a bearing only

True bearing, 018°

Mark, transit with view

Historical fort

Quay, berth dries *4.6m* One side inaccessible

Amer

Balise

HAT

LAT i.e. Chart Datum

3 metre contour

Chapel, Church

Navigationally important rock in a drying area, dries *2.1m*

Buildings

Lifeboat station and slip

Slip, end dries *1.3m*

Spot height of 25m above MHWS

Water-tower

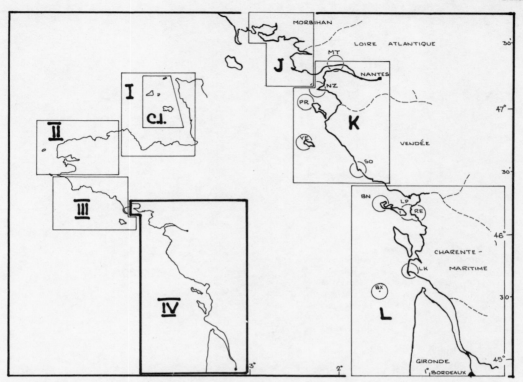

Fig. 2. Index to book sections

big difference between French and British charts is in the height of rocks which never cover, but I've followed the Admiralty by showing heights as measured above MHWS. On modern French charts these heights are measured from 'mean level' or about half tide, although on older ones they are shown as the height above chart datum, very confusing. Headway under bridges is the same and I give this as above MHWS. Bearings are true; those of lights from seaward; of transits as seen in the view. Symbols are generally those on Admiralty Chart 5011, alas without colours, Fig. 1. Section that down the middle and you have approximately Fig. 12. I very seldom show rocks—the area is wall-to-wall anyway—unless of significance. Their elevations I've guessed sometimes to point the difference between 20m and 5m; who cares if they really should be 17·8 and 5·3? Drying ground may contain rocks, sand, shingle, mud, or a mixture of the lot. Similarly the quality of the bottom is ignored; however no anchorages are shown in known poor holding.

Compass roses are absent, the side margins are true north. East margins show minutes of latitude and tenths, i.e. nautical miles and cables. The south gives minutes and tenths of longitude. Where these are omitted I draw a scale. A cable = 185 metres = 608 feet. To give some clue to the terrain, spot heights are sometimes shown in metres above MHWS.

Stupid mistakes are inevitable in constantly changing from British to French charts; swapping from metres to feet; confusing abbreviations like B = black = *blanc*. Therefore my

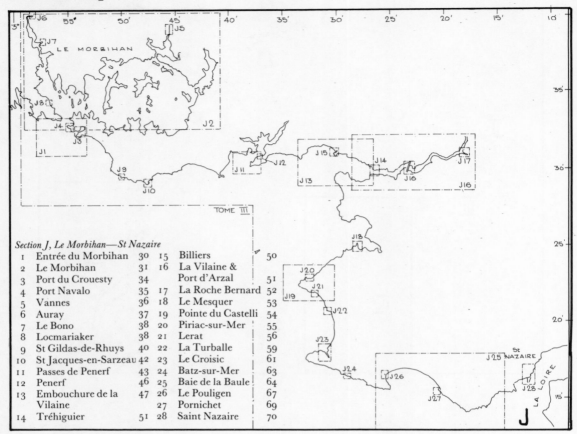

Fig. 3. Index J. Le Morbihan to St Nazaire

views and charts, abbreviations, colours, lights are all in French. My charts are never intended to substitute for proper navigational charts. But when you buy these, buy French because Admiralty coverage is meagre and the scales are small. Some 15 charts cover the same area as 40 French ones, which can be bought from most Admiralty agents in the U.K. and in many French ports. Keep them corrected from Notices to Mariners, clean, dry and flat for they cost double; Admiralty charts are indexed in Fig. 6, French small-scale passage charts in Fig. 7. Larger-scale detail French charts are shown in Fig. 8.

A single word, light, describes lighthouses, light beacons, light structures; characteristics, which often change, are never given, only height above MHWS; sectors are occasionally referred to. In thick weather or gathering dusk it is comforting to recognise quickly a beacon, light, which is why I have included so many of their profiles. A *balise* is a pole with a topmark, but if it's on a rock it might have a concrete base for strength. A *tourelle* is of masonry or concrete, commonly like a capsized flowerpot and always with a topmark. A disused *tourelle* is painted white. It is promoted to a *phare* (lighthouse) if it is given a light, indicated by a star. *Amers*, many of which are relics of an era before lights and uniform buoyage, are daymarks, more often white than coloured and very often form one of a pair of transit marks. They never

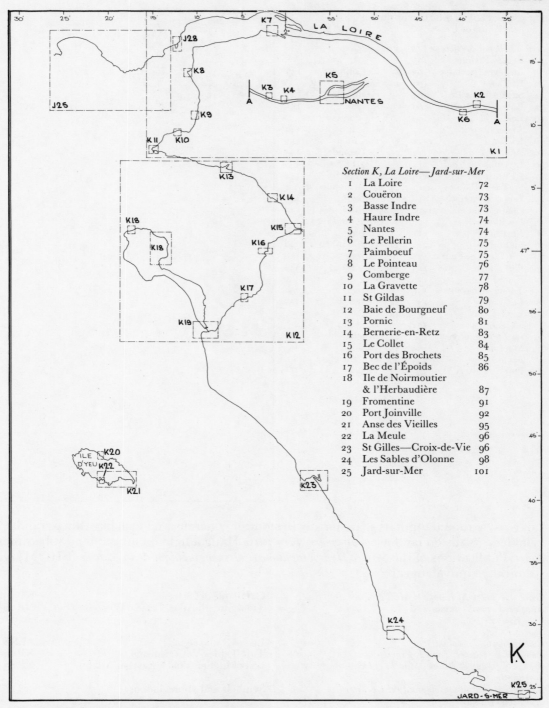

Fig. 4. Index K. La Loire to Jard-sur-Mer

French Pilot 4

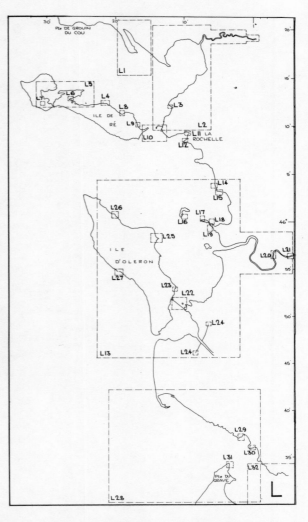

carry navigational topmarks, are mostly in stone or concrete, and look like slim pyramids or cylinders. Walls on land are *mur-amers*, very rare. Hang a light on an *amer* and it becomes a *phare*. Publications of the *Service Hydrographique et Océanographique de la Marine* (SHOM) with Admiralty equivalents are:

Catalogue-Index A, Europe et Méditerranée	4A	Catalogue of Charts	NP 131
Atlas des Courants de marée,		Tidal Stream Atlas, France West coast	NP 265
de Brest à			
St Jean de Luz	552		
Radiosigneaux (1ᵉʳ volume)	91	List of Radio signals	ALRS (2)
Annuaire des marées, Tome I, France	5	Tide Tables Vol I, Europe	NP 200
Feux et Signeaux de Brume, Manche et Océan		List of Lights, Vol D, Eastern Atlantic	NP 77
Atlantique est	C		
Symboles et abréviations figurant sur les cartes		Symbols and abbreviations	5011
marines françaises	1D		
Instructions Nautiques, France nord et ouest	C2	Biscay Pilot	22

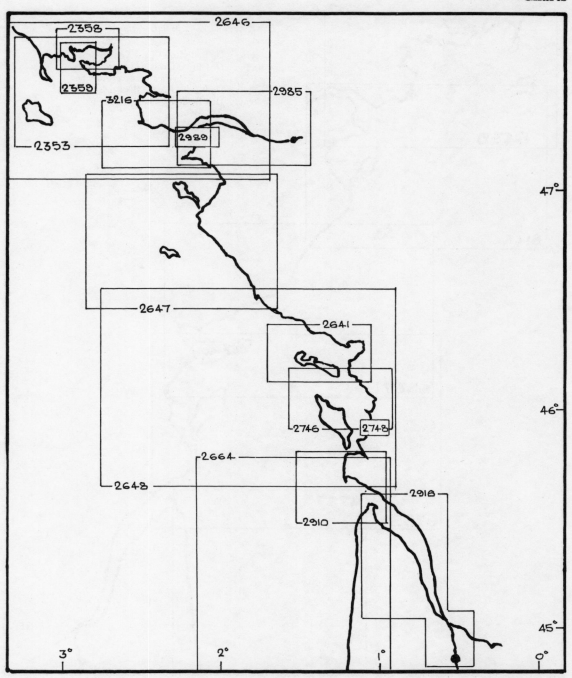

Fig. 6. British Admiralty chart index

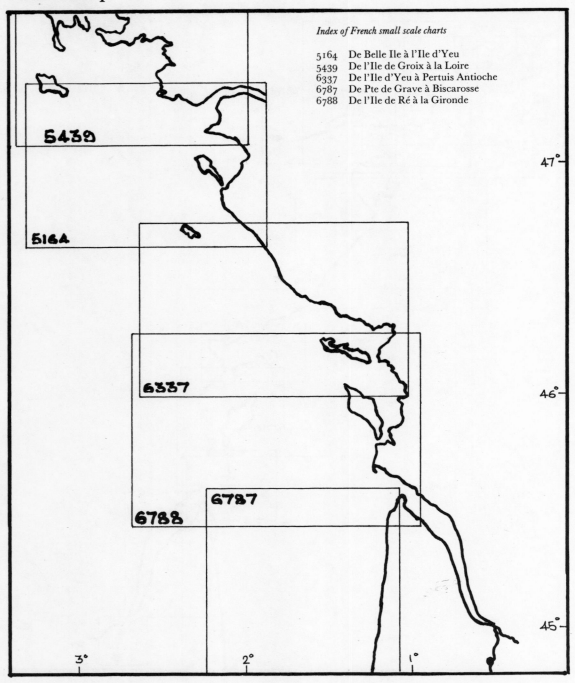

Index of French small scale charts

5164	De Belle Ile à l'Ile d'Yeu
5439	De l'Ile de Groix à la Loire
6337	De l'Ile d'Yeu à Pertuis Antioche
6787	De Pte de Grave à Biscarosse
6788	De l'Ile de Ré à la Gironde

Fig. 7. French index of small scale charts

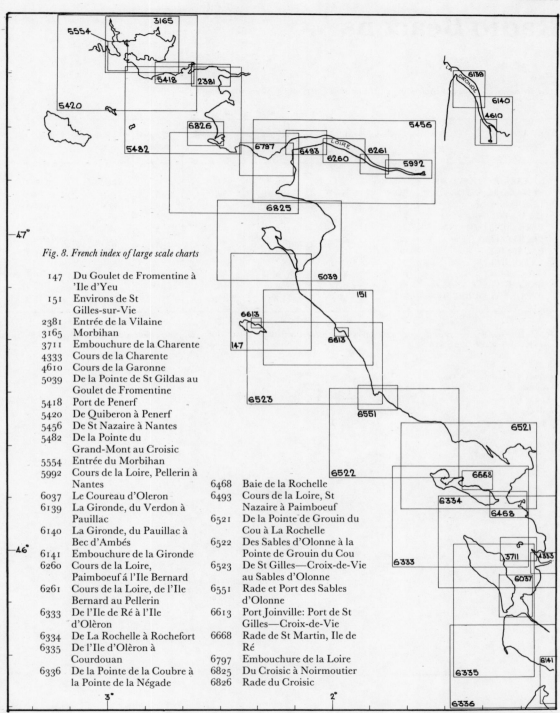

Fig. 8. French index of large scale charts

Fig. 8. French index of large scale charts

Radio Beacons

These are shown on Fig. 2 as a circle with the identification letters inside. All are continuous.

Name	Type	Ident.	Frequency kHz	Emission	Range miles	Position N	W
ST NAZAIRE/MONTOIR	Aero RC	MT	398.0	A1	50	47°20.0′	2°02.6′
ST NAZAIRE, GILDAS (1)	RC	NZ	289.6	A2	35	47°08.1′	2°14.7′
ILE DU PILIER	RC	PR	298.8	A2	10	47°02.6′	2°21.5′
ILE D'YEU	RC	YE	312.6	A2	70	46°43.1′	2°22.9′
LES SABLES D'OLONNE	RC	SO	291.9	A2	5	46°29.6′	1°47.7′
LES BALEINES (2)	RC	BN	303.4	A2	50	46°14.7′	1°33.6′
LA PALLICE	RC	LP	287.3	A2	5	46°09.7′	1°14.3′
LA ROCHELLE, RICHELIEU	RC	RE	291.9	A2	5	46°08.9′	1°10.4′
BXA LIGHT FLOAT	RC	BX	291.9	A2	5	45°37.6′	1°28.6′
PTE DE LA COUBRE (2)	RC	LK	303.4	A2	100	45°41.9′	1°13.9′

Several of the above stations are grouped in the following sequences with the beacons listed, at one-minute intervals. Full details are in Admiralty List of Radio Signals—Vol 2 ALRS (2)

(1)		(2)	
Eckmühl	ÜH	Belle Ile	BT
St Nazaire	NZ	Ile de Sein	SN
Pte de St Mathieu	SM	Pte de la Coubre	LK
Pte de Combrit	CT	Les Baleines	BN
Ile de Groix	GX		

Marks

Alignments, transits, lines, marks, they all mean one thing—a pair of objects one behind the other—which is all this book is about. Inshore pilotage depends entirely on marks, which word also describes a church, an *amer*, a house. A channel mark is when an alignment passes between dangers on both sides. A clearance mark keeps you away from dangers on one hand. A breast mark is secondary to a main mark, is roughly abeam, and usually tells when to turn, i.e. quit one mark and take another. How much to deviate from a mark, yet still be safe can only be judged from charts giving enough detail. Occasionally I have stressed one side or the other. The distant mark is given in the text before the nearer one; the sign '×' means 'by' to avoid confusion. A typical view is in Fig. 9—line 12Z a tower on a hill × a beacon on a rock. Sometimes the marks are not in dead alignment as when the dotted rear mark is opened. In Fig. 9 it is opened to the SW but to avoid any mistake I would write 'a tower to the right of a beacon on a rock'. This is safest when required instantly to change from an ahead to an astern transit. When going along using a stern mark, it is pointless and dangerous to look

Fig. 9. Typical transit view

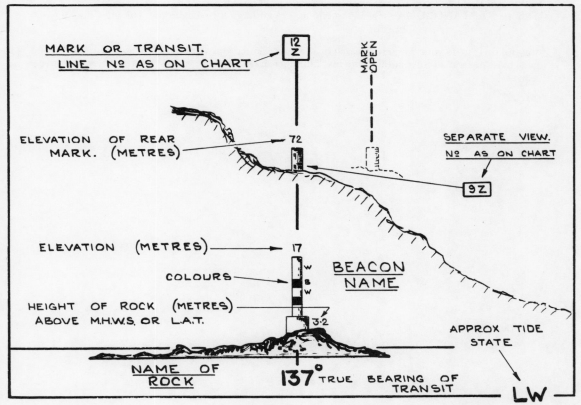

ahead only. Make a check for lobster pots, other boats, etc., then turn your head and concentrate on the transit. Have ye no faith in my marks?

All the marks have been sailed or motored over, according to tide, in our 44 foot sloop *Hephzibah*; she draws 2·10 metres and from her decks, her cockpit, her crosstrees I have made the sketches. May I therefore quickly agree with all you art critics? But I'm a sort of seaman and not an artist; try struggling with wet paper, one hand on the chart and another on a bearing compass, binoculars under your arm, half an eye on the sounder, bellowing into a tape recorder—all at once. Then repeat the whole lot because it's too late for the tide, again since there's too much swell, or it's neaps instead of springs, better scrub it for today anyway for it's too misty to see the marks!

Perspective has been fiddled and detail omitted, both for clarity. If an object is framed, this shows that it is out of position. If of use I have sometimes shown the approximate tide level, but only LW, HT and HW, (BM, MA, PH) on average tides. Numbering of the lines is easy and I've tried to keep them in sequence. Charts are prefixed by the letters of their section (J1, J2, J3). Views in that section are suffixed by the same letter (25J, 26J, 27J). There are TWO kinds of transit numbers:

Within a SQUARE — A view with a transit on a chart, of the same number.
Within a RECTANGLE—A view of something, drawn large for recognition. The view may have a bearing.

A useful tip is to do your 'homework' well in advance of entering an unknown harbour. Draw, on your own chart, my transit lines, adding the mark number, page reference of its view together with its true bearing.

Safety

France looks after her citizens who take to the sea by a system of licences for boats and examinations for skippers. There are some 20 weather forecasts daily (more in summer) and a list of times, frequencies and automatic telephones is given in a free leaflet *La Meteo* from any *Douane*, *Bureau du Port* or *Inscription Maritime*. The leading shipping forecasts, though times are liable to change, are

St Nazaire (1722 kHz)—0703, 1703 GMT in French.
Bordeaux–Arcachon (1820 kHz)—0603, 1603 GMT in French.
BBC (200 kHz)—0015, 0625, 1355, 1750 CLOCK times.

The equivalent to coastguard stations are semaphores. These show storm warnings and are at prominent points:

> St Nazaire, Pte de Chémoulin
> Ile d'Yeu
> Les Baleines
> Oléron, Pte de Chassiron
> Pte de la Coubre

The headquarters of the French lifeboat service for Biscay is C.R.O.S.S.—A, Château de la Garenne, 56410 Étel: Tel (97) 52.35.35. Lifeboat stations, 24 hour watch, are at

> l'Herbaudière
> Ile d'Yeu
> St Gilles—Croix-de-Vie
> Les Sables d'Olonne
> La Pallice
> La Cotinière
> Port Bloc

Marks and Transit Lines

French Pilot 4

Marks & Transit Lines – Continued

Buoyage

The promised change from the previous excellent French system to the new IALA buoyage is now complete in Biscay. This fresh look at maritime marking devised by the International Association of Maritime Authorities is that now used throughout Europe and the daytime marks are shown in Fig. 13.

Except for lighthouses and leading marks it is used for all fixed and floating marks. Objects are painted in four colours, lights have three. The five types are unambiguous:

LATERAL marks are on the sides of channels, the sense being that when coming in from seaward.
CARDINAL marks show where navigable water is in relation to, i.e. FROM a danger. Note that the French *ouest* (west) is abbreviated to the English W. CW means Cardinal West etc.
ISOLATED DANGER marks are built on or moored over a danger with navigable water all round.
SAFE WATER buoys are for mid-channels and landfalls.
SPECIAL marks are for military areas, traffic separation, outfalls, etc.

Whether you are looking at a *balise*, buoy, *tourelle* or pierhead, it is the topmark and, to a lesser extent, the colour which are the clues. That is why I have drawn, in Fig. 13, the dotted shapes, which vary considerably in practice. Abbreviations:

Blanc	b	White
Aux éclats	é	Flashing
Fixe	f	Fixed
Jaune	j	Yellow
Aux occultations	o	Occulting
Rouge	r	Red
Scintillant	sc	Quick flashing
Scintillant rapide	sr	Very quick flashing
Vert	v	Green

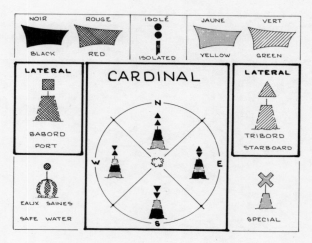

Tides

Unlike the areas covered in my three previous Pilot books (Normandy, Channel Is and Brittany) the Bay of Biscay has neither the large tidal range nor, in consequence such fierce currents. So I have scrubbed tidal flow diagrams and instead give you in Fig. 10 the strength and direction of currents at selected points. Except at these places and at river mouths the tidal currents along the coast, within this volume, elsewhere seldom exceed 1½ knots.

At any port in France you can pick up, free from any newsagents, a local tide table. This gives the day's times of high water (PM) and low water (BM). Alongside each is a figure, a 'coefficient' which is proportional to the range of tide, morning or evening, on that particular day.

Coefficients vary between 45 at neaps (*Morte eau*, ME) and 95 at springs (*Vive eau*, VE) though for equinoctial tides they can be as high as 120 and as low as 20. The table in Fig. 11 shows the heights of high and low water, springs and neaps at several well spaced places; you will see that there is little variation. Though there is a rather complicated way of calculating more exactly, it is really quite accurate enough to estimate for those days when the coefficient differs significantly from 45 or 95. Thus for a coefficient of, say, 70 (half way between 95 and 45) you could take a height half way between the springs and neaps figures.

Fig. 10. Tidal currents

Port of reference

	St Nazaire Mouth of the Loire N 47°15′ W 2°13′			La Rochelle Pont d'Oléron N 45°52′ W 1°11′			Cordouan Pointe de Grave N 45°35′ W 1°3′		
Hours	direction degrees	springs knots	neaps knots	direction degrees	springs knots	neaps knots	direction degrees	springs knots	neaps knots
−6	218	1·8	0·9	022	1·0	0·5	330	1·0	0·6
−5	033	0·3	0·1	—	—	—	143	0·5	0·3
−4	045	2·5	1·2	180	0·5	0·3	140	2·2	1·3
−3	043	2·7	1·4	—	—	—	143	3·7	2·2
−2	048	1·8	0·9	180	1·5	0·8	145	3·2	1·9
−1	080	1·0	0·6	—	—	—	144	2·6	1·6
High water	140	0·3	0·2	180	1·5	0·8	148	2·3	1·4
+1	216	0·8	0·4	—	—	—	165	1·5	0·9
+2	222	1·2	0·6	180	1·0	0·5	235	0·9	0·5
+3	224	3·0	1·7	—	—	—	308	2·5	1·0
+4	221	3·1	2·2	022	0·5	0·3	315	3·6	2·2
+5	217	2·7	2·1	—	—	—	315	3·7	2·2
+6	219	2·3	1·5	—	—	—	323	2·7	1·6

Fig. 11. Tidal heights

	springs		neaps	
	MHWS	MLWS	MHWN	MLWN
Port Navalo	4·9	0·6	3·8	1·7
St Nazaire	5·2	0·5	4·1	1·7
Ile d'Yeu	4·9	0·5	3·8	1·7
La Rochelle	6·0	0·8	4·8	2·4
Pointe de Grave	5·4	1·0	4·3	2·0

Tidal definitions are explained below and are graphically shown in Fig. 12.

HAT (Highest astronomical tide) = **PHM** (*plus hautes mers*) and **LAT** (Lowest astronomical tide) = **PBM** (*Plus basses mers*) are the levels which can be predicted to occur under average meteorological conditions and under any combination of astronomical movements.

MHWS (Mean high water springs) = **PMVE** (*Pleine mer de vive eau*) and **MLWS** (Mean low water springs) = **BMVE** (*Basse mer de vive eau*) are the average rises throughout the year on two successive tides when the moon is at 23½° declination and the tide is greatest.

MHWN (Mean high water neaps) = **PMME** (*Pleine mer de morte eau*) and **MLWN** (Mean low water neaps) = **BMME** (*Basse mer de morte eau*) are for the same conditions as above but when the tide is least.

Just one reminder . . . BST, GMT or French time, but not a combination of all three. So when using tide tables remember that French time is one and sometimes two hours different from British.

Fig. 12. Tidal definitions

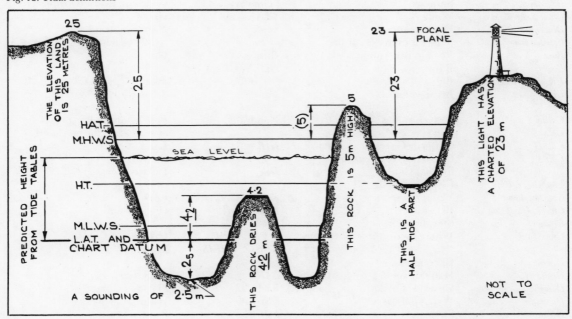

Glossary

Only words in the text and on charts are given. Breton words are in capitals. In this Celtic language letters are frequently interchanged, i.e. G–K–C'H, S–Z, G–W, B–V–P. An interesting dictionary of Breton– French is published by: Librairie de Finistère, 51 rue du Château, Brest 29N.

ABER	River mouth	Dog-leg, zig-zag	Two courses parallel, but offset, staggered
About	E.g. south-about is to set a course to leave an object to the north	*Dragué*	Dredged
Amer	Daymark	*DU*	Black
Ancien, -nne	Disused, old	Duc d'Albe	Mooring dolphin, pile
Anse	Small bay, cove		
AR, AN, AL	The	*Écluse*	Lock
AVEN	River	*EL, EN, ER*	In the
		ENES, ENEZ	Island
Balise	Perch, pole beacon	*Épi*	Spur of a quay
BANN, BENN	Height		
BAS, BAZ, BATZ	Shallow	*Feu*	Light
Basse	Low rock	*FEUNTEUN*	Fountain
Bec	Point	*Fosse*	Ditch, channel
BEG	Point	*FROUDE*	Rapids
Beniget, beniged	Gap		
BENVEN, BOSVEN	High rock, stack	Gare SNCF	Station, French railways
BIAN, BIHAN	Little	*GARO*	Red deer
Bili, vili	Shingle	Gateway	Course between two objects
Blanc, -che	White	*GLAS*	Green
Blanchi	White painted	*Goulet*	Narrow entrance
Bouchot à moules	Stakes for mussel culture	*Grève*	Sandy beach
Boue	Rock mostly submerged	*Gris, -e*	Grey
Bouée	Buoy	*GUEN, GUENN,*	
BRAS, BRAZ	Big	*GWEN*	White
Brise-lames	Breakwater	*Guérite*	Small watch tower
Cailloux	Gravel, stones	Handrail	Passing around an object,
Cale	Slip		or several on the one hand
C de G	Corps de Garde, watch house	*HIR*	Long
C de S	Canot de Sauvetage, lifeboat station	*Isolé*	Isolated
Chapelle, église	Chapel, church		
Château d'eau	Water tower	*KARREG, CARREC*	Rock
Cloche, Clocheton	Bell, turret	*KER*	House, hamlet
Clocher	Steeple, belfrey	*KREAC'H,*	
Crossroads	Where two or more marks meet	*CREAC'H*	Hillock
		KREIZ, CREIS	Middle
Damier	Chequered		
Demie	Half tide rock	*LANN*	Monastery
Déroché	Cleared of rocks	*LEAC'H*	Flat stone, place
Déversoir	Weir, spillway	*LEDAN*	Wide
Digue	Stone dyke	*LOC'H*	Pond, mere

MARC'H	Horse	*Raz*	Race
Marégraphe	Tide recorder	*Robinet*	Tap
MEAN, MEN	Stone, rock	*Roche, ROC'H*	Rock
MELEN	Yellow	*Rocher*	High rock
MENEZ	Hill	*Rouge, RUZ, Rousse*	Red, reddish
Menhir	Standing stone	*ROZ, ROS*	Wooded
MERC'H	Daughter, girl	*Ruisseau*	Rivulet, stream
Méridional, -e	Southern		
MEUR	Great	*Sal Br.*	Signal de brume, fog signal
MEZ	Seaward, wide	*Seuil*	Cill, sill
MOR, VOR	Sea, salt water	*Sifflet*	Whistle
Mouillage	Anchorage	*Son*	Blast
Moulin	Windmill, mill	*Sonde*	Sounding
Musoir	Pierhead	*STER*	River, creek
Neuf, -ve	New	*Terre plein*	Levelled area near quay
NEVEZ	New	*Tête*	Head
Nez	Nose	*Tirant d'air*	Headway, clearance
Noir, -e	Black	*Tirant d'eau*	Draught
Nouveau, -elle	New	*TOULL*	Cave, hole
		Tourelle	Navigational tower
Occidental, -e	Western	*TRAOU*	Low, west, valley
Oriental, -e	Eastern	*Traverse*	Intersecting channel
		TREAZ, TREZ	Sandy
PELL	Distant	*TREIZ*	Passage
PEN	Head, headland		
Phare	Lighthouse	*UHEL*	High
Pierre	Stone		
Pignon	Gable	*Vanne*	Sluice gate
Plat, -e, tte	Level, flat	*Vert*	Green
PLO, PLOU, PLU	Parish	*Vieux, Vieille*	Old
PORS, PORTZ, PORZ	Harbour, inlet	*VIR*	Needle
POUL, POULL	Roadstead, pool, lagoon		
Presqu'île	Peninsular		

General Information

Customs. Pleasure boats arriving in France from abroad may only enter at a port with a *douane,* must have ships' papers, crew list and stores inventory, must fly international day or night signals. With whimsical Gallic logic customs go on to say that by 'tacit declaration' and provided you arrive by sea, don't trade, run immigrants or charter to French citizens, you are free to come and go as you please. They can, and sometimes do, take a random check and can board your boat within 20 miles of the shore. The system is under threat of change.

Immigration. Passports only, required but seldom asked for, unless quitting the country by some other route. You will need them for cashing cheques and for *poste restante* mail.

Health. Certificates not required. If you need medical treatment in France see the Foreign Office leaflet 'Essential information for U.K. passport holders'.

Pets. Crippling fines if you bring a mammal back to the U.K. and practically a death penalty in the Channel Islands.

Money. Credit cards everywhere and all banks cash travellers cheques with no limit—or, with a Eurocard, cheques on U.K. banks. You mustn't take away more than 5000 Frs in notes.

Security. I've never locked my boat in any part of France and see no reason to start.

Duty Free. Can be bought wherever there is a willing chandler in a customs port.

Speed limits. 5 knots under motor within 300m of the shore.

Mopeds. Visitors, whose mopeds under 50 cc are licensed in their own EEC countries, may use them in France without further documentation as long as the riders wear crash helmets, keep off motorways and use cycle ways where available.

Acknowledgements

There are two reasons why this book doesn't pretend to be a literary gem. For a start it's mostly pictures, isn't it? And though it has been a most pleasant exercise in communication, there is nothing in it which is original; other folks' advice, other people's ideas. Pilots have helped, commercial skippers have cautioned, yachtsmen have confessed secrets; here a lead, there a kindly chunk of advice. I can't think of an example when I was refused help. I've already drunk their health on board, but here is the place to shout *Yëa-a-mad* to all those who joined in the research ... Myrtle Green, Beth and Dennis Hurden, Captain Louis Lecoublet, John Lintell, Ann and John Moorshead, Bill Pethick, Freddie Torode, Commandante Richard Winter and my wife, Joan (who also co-researched much). Oh yes, and don't let us forget dear old *Hephzibah*, who looked after us all so well—she looks a shade tired after a non-stop flit in and out of some 180 harbours.

My sketch charts are based on French charts with the sanction of M. l'Ingénieur Général de l'Armement EYRIES, *Directeur du Service Hydrographique et Océanographique de la Marine*. I am truly grateful for his Department's help.

J Le Morbihan to St Nazaire

When sailing down the Biscay coast of France, you could say that the general character changes about where this book starts. Cliffs, valleys and headlands dominate the rugged coast as far as the river Loire, but now we have in front of us some 250 miles of gentler scenery, wide, shallow bays and low islands and two of the largest rivers which flow into the Gulf. It is here, where the rocks end, the wine begins.

The section starts off with a chunk of coast rightly belonging to the tail end of Lower Brittany; two departments, Morbihan and Loire Atlantique. Le Golfe du Morbihan gives way to the country of Guérande which ends at La Grande Briére on the west of the Loire river.

LE MORBIHAN

The only way to explore the scores of islands, the limitless anchorages and the tiny quays of this vast inland sea is by water. SHOM chart 3165 covers the whole area, so my chart J2 is diagrammatic. To be practical, I have selected the most important harbours only. There is a score of anchorages shown on the chart. Note that the place names, *balises*, buoys

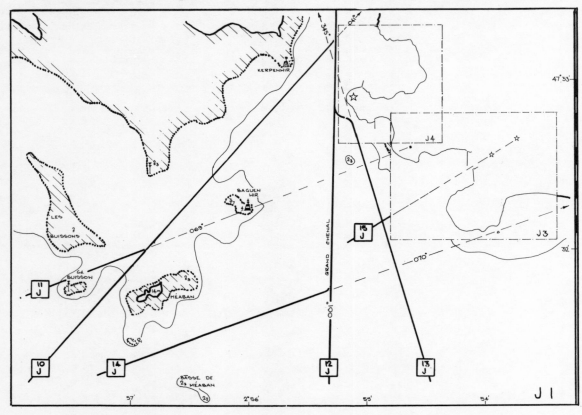

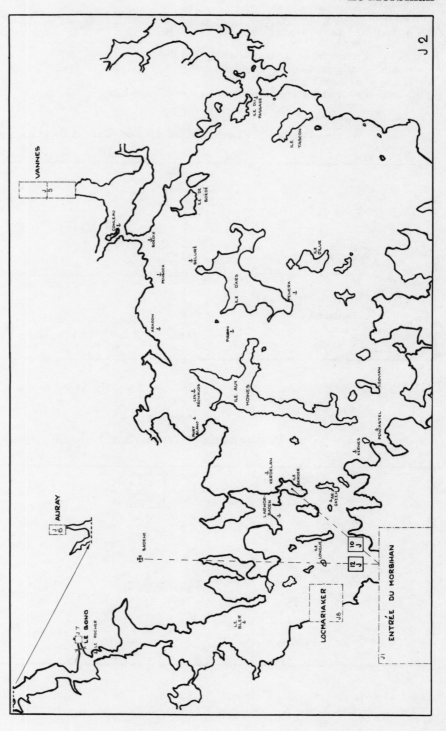

and soundings are not shown. Some care is needed with the entrance, chart J1, but the main deep channels are well marked:

    1—Passage between Méaban and Grand Buisson
    2—Passage au nord du Grand Buisson
    3—Le Grand Chenal
    4—Passage between Méaban and La Basse de Méaban

1—Passage between Méaban and Grand Buisson (0·5m)

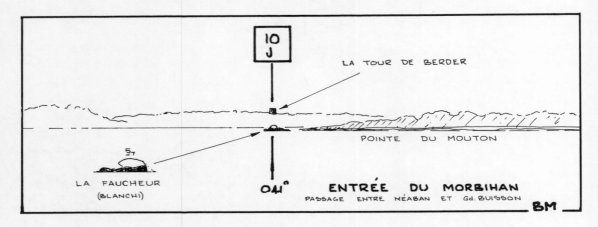

10J La Tour de Berder × a whitened rock off Pointe du Mouton, La Faucheur.

2—Passage au nord du Grand Buisson (2·8m)

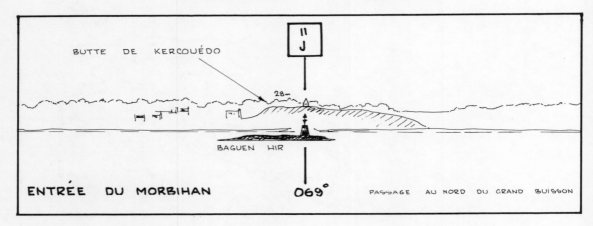

11J Butte de Kercouédo × Baguen Hir *tourelle*. The Butte has a small house near its peak, and this mark joins the preceding one to the north of Méaban.

3—Grand Chenal (7·0m)

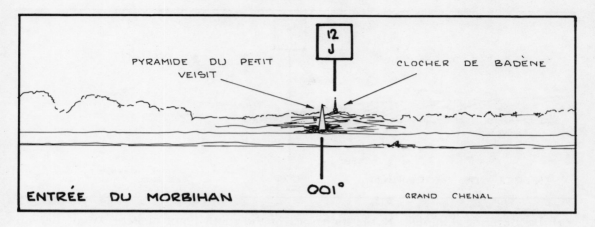

12J Badène church × a white *amer* on the Ile du Petit Veisit. As a tide-cheating alternative to the above, try

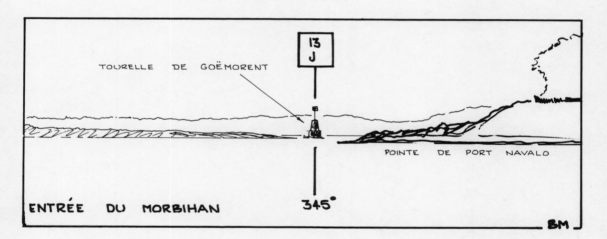

13J Goëmorent *tourelle* × Pointe de Port Navalo. This mark takes you inside the Basse du Talec and can therefore be used either for Crouesty or Port Blanc.

4—Passage between Méaban and le Basse de Méaban (4.4m)

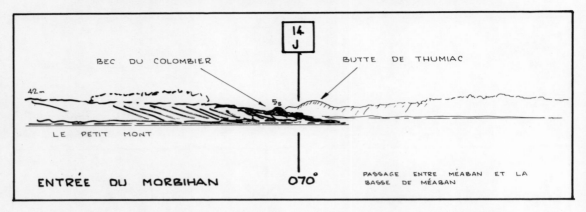

14J La Butte de Thumiac × the high rock of the south point of Petit Mont, Bec du Colombier.

Port du Crouesty

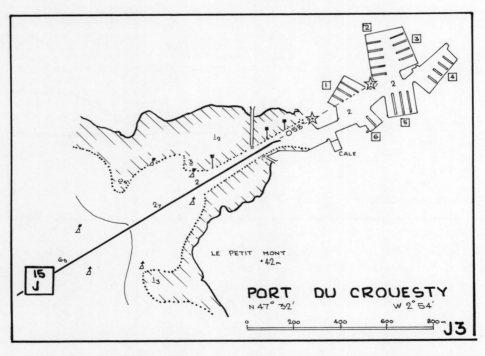

Just before tangling with the fierce tides of the Morbihan entrance, there is a large new marina, with places for 1,000 boats and for a further 100 visitors, chart J3. The day or night marks are

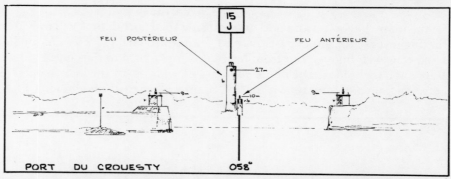

FEU POSTÉRIEUR   FEU ANTÉRIEUR

27m

10m

PORT DU CROUESTY   058°

15J the two lighthouses in line. The channel is dredged 2m, well buoyed and this modern marina offers every facility.

Port Blanc

Sometimes on the ebb in the entrance to the Morbihan, you feel like a salmon trying to crawl up a waterfall, so why not catch your breath at Port Blanc, chart J4? You can dry out comfortably on the east side of the slip and the shops at Port Navalo are but a kilometre away.

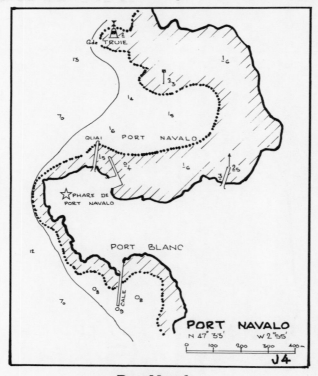

PORT NAVALO
N 47° 33'   W 2° 55'
J4

Port Navalo

This is a most convenient drying harbour and a kind of waiting-room for the Morbihan, chart J4. The quays must be kept free for the constant ferries from Vannes and Locmariaker, but there are sundry buoys in the non-drying part in the middle.

Vannes

Of the many harbours and odd quays inside the Morbihan, I will only describe four of which the biggest is the important town of Vannes, chart J5. The channel as far as Pointe de Larmor is well buoyed but from here on, simply keep to the centre. The ferries use the Pont Vert but a toehold can be found there while waiting for the lock to open. The lock is opened, between 8.00 a.m. and 9.00 p.m., 2½ hours before to 2½ hours after high water.

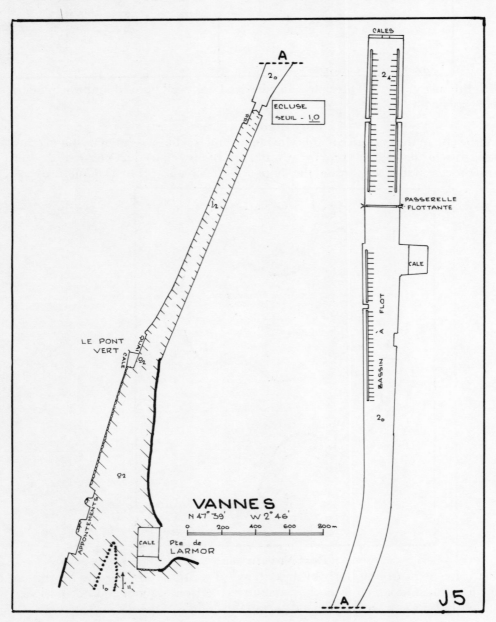

VANNES
N 47°39'   W 2°46'

0   200   400   600   800 m

Auray

Far less commercial, this lovely medieval town has drying quays a-plenty. Beware of the Quai Franklin (named after the American inventor), because it has dangerous arches exposed at half-tide. Chart J6.

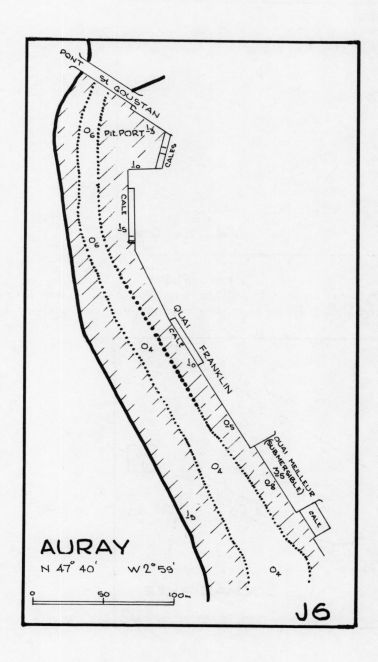

AURAY

N 47° 40'   W 2° 59'

0   50   100m

J6

French Pilot 4

Le Bono

I'm afraid I can't give you the headroom under the motorway bridge, though ample for most small craft, which you must pass just before getting to Le Bono, chart J7. You can use either side of the quay or the extensive, though muddy, basin.

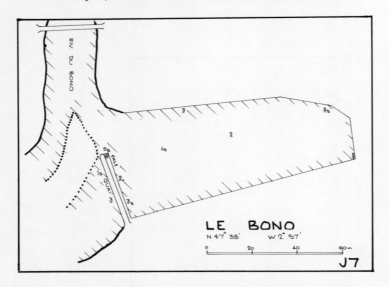

LE BONO
N 47° 38'   W 2° 57'

Locmariaker

A largish village, served by ferry from Vannes and Port Navalo, chart J8, it has the largest *menhir* in Brittany, Men-er-Hroëc'h. There are two channels through the mud. The first is

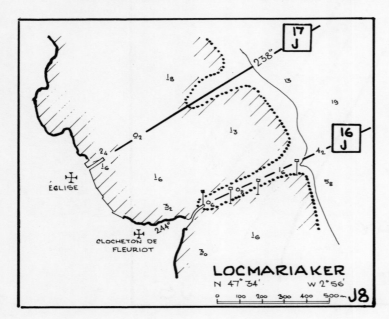

LOCMARIAKER
N 47° 34'   W 2° 56'

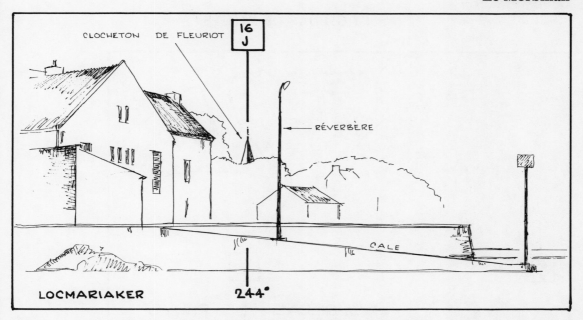

CLOCHETON DE FLEURIOT

16 J

RÉVERBÈRE

CALE

LOCMARIAKER

244°

16J Fleuriot spire seen to the right of a large white house at the top of the slip. This slip is in constant use by the ferries. Much more tranquil is the quay opposite the church which, since it dries rather a lot, is seldom used. Take

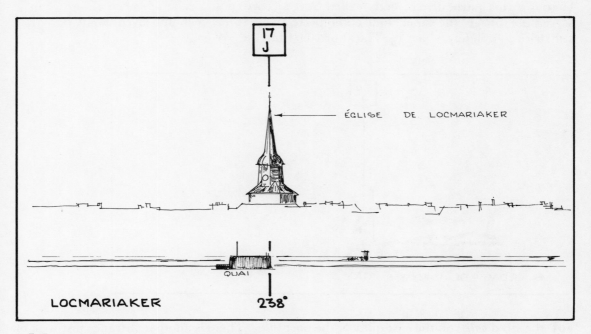

17 J

ÉGLISE DE LOCMARIAKER

QUAI

LOCMARIAKER

238°

17J Locmariaker church ✕ the end of the quay.

ST GILDAS-DE-RHUYS

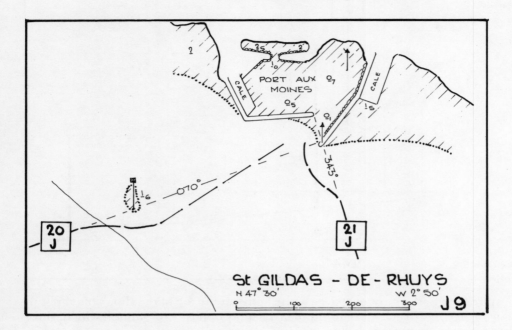

Here is a tiny harbour, correctly called Port aux Moines, a relic of earlier times when it served the town of the same name a kilometre away. It has been improved lately, though it still dries completely, chart J9. The approach from the SW is

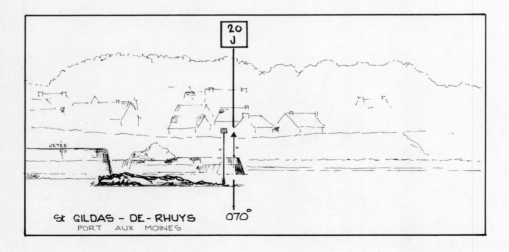

20J the stb'd *balise* on the east quay × the port *balise*. There is another entrance just west of Le Bauzec *tourelle*,

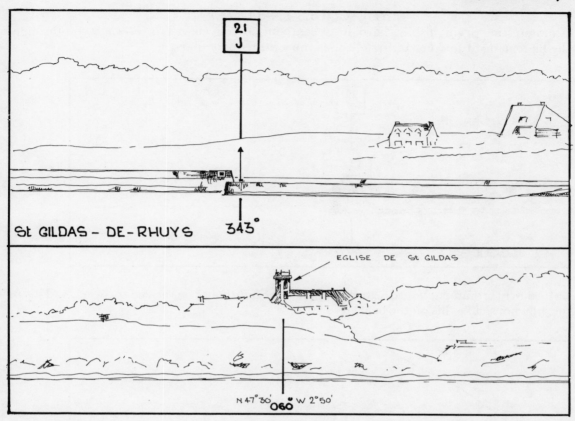

21¹
J

St GILDAS – DE – RHUYS   343°

EGLISE DE St GILDAS

N 47° 30'   **060**   W 2° 50'

21J the red patch on the west quay ✕ the stb'd *balise* on the east quay. On the same sketch there is a portrait of the nearby very prominent St Gildas church.

ST JACQUES-EN-SARZEAU

An interesting drying fishing harbour, recently improved, chart J10. To clear you through the Plateau de St Jacques I can only offer you a rather poor mark,

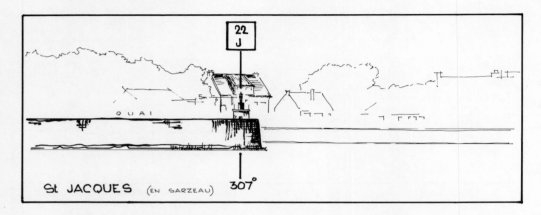

22J a fairly prominent house near the beach × the lighthouse on the end of the quay. This is the only house like this on the bearing.

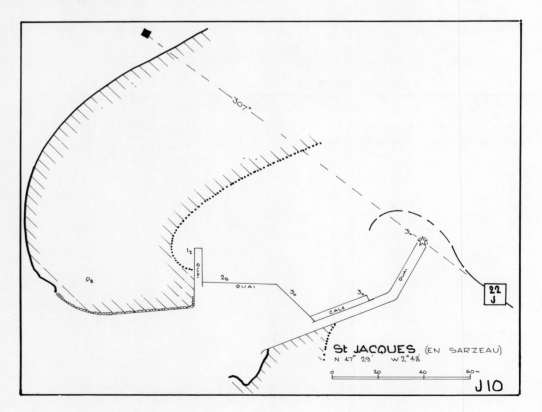

PENERF

Fortunately, this river has a difficult though unchanging entrance which makes it considerably quieter than many harbours. There are three entrances, chart J11:

      1—Passe de l'Ouest
      2—Passe de l'Est
      3—Passe du Pignon

The channels end at the slip of Penerf which is 1½ miles from the entrance light. For want of names I have given all the *balises* letters A–H. A very valuable recognition object is the Tour de Penerf whose portrait is on view 34J, (p.48).

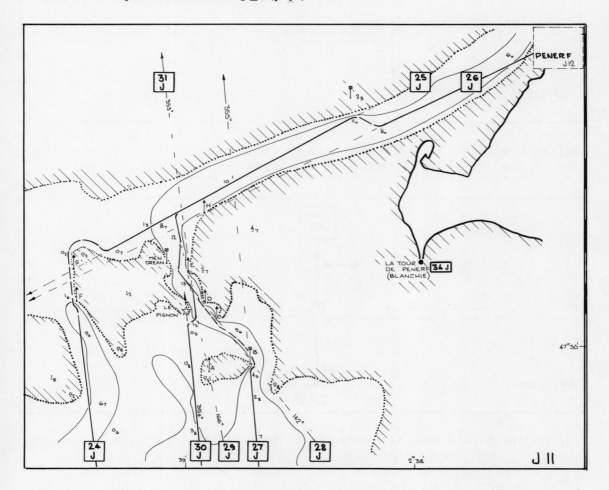

1—Passe de l'Ouest (0·2m)

This starts a couple of cables west of the Plateau de Penvins port buoy.

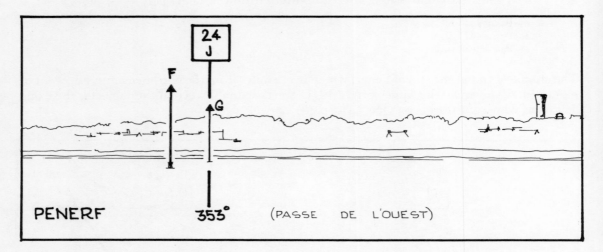

24J Two stb'd *balises* 'G' and 'F' open as shown. When within spitting distance of the near one, 'F', pass both 'F' and 'G' 50m off and make a handrail to the north of 'G' until

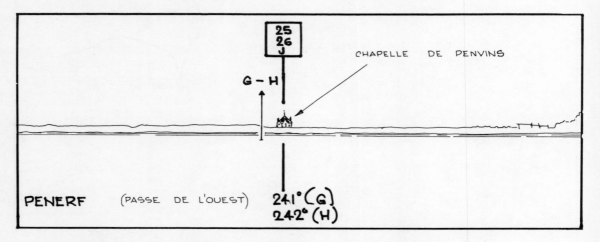

25J Penvins chapel × *balise* 'G'. Continue until abreast of the first port *balise*, and using the same sketch above, come to the SE until

26J the same Penvins chapel × another stb'd *balise* 'H' which takes you right up to the slip in Penerf.

2—Passe de l'Est (2·8m)

This is the deepest though most complicated of the three channels. You can see on the north skyline two prominent objects, a water tower and a church spire. The latter is our mark,

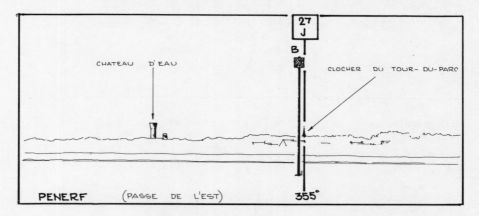

27J Tour-du-Parc spire × port *balise* 'B'. When 50m from the *balise*, leave it 30m to port and steer towards stb'd *balise* 'C'. Though without transits, there is plenty of water in this stretch. Leave 'C' and 'D' 30m to stb'd and take a back mark,

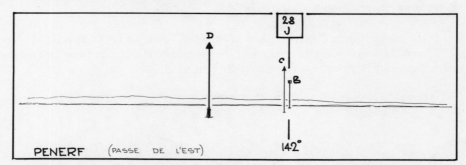

28J *balise* 'B' × *balise* 'C'. When abreast of 'E' turn to stb'd to look for a back mark,

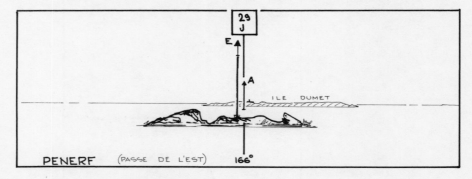

29J two stb'd *balises* 'A' and 'E' in line which joins 25J (p.44).

3—Passe du Pignon (0·5m)

The first mark is simple,

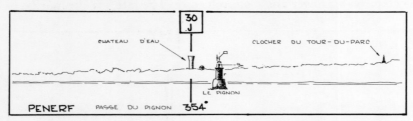

30J a water tower to the left of Le Pignon lighthouse. When 100m off, steer to stb'd until

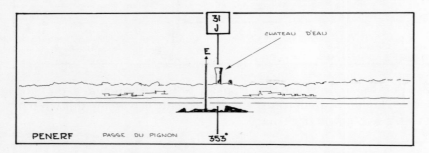

31J the water tower is to the right of *balise* 'E'. After a cable this joins 28J (p.45).

There is plenty of anchor room in the river, chart J12, and one or two moorings; and with care you could dry out on the slip near Penerf chapel.

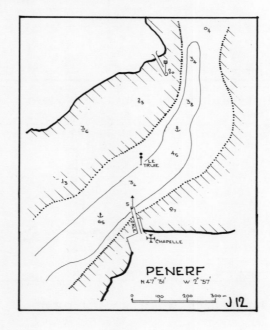

LA VILAINE

Here is another completely unspoilt river but with an unfortunate bar at the entrance. However this does not move and with the aid of tide tables, the marks, which are quite ancient, can be followed with confidence, chart J13.

There are three channels:

1—Chenal de la Grande Accroche
2—Passe de la Varlingue
3—Chenal Est de la Varlingue

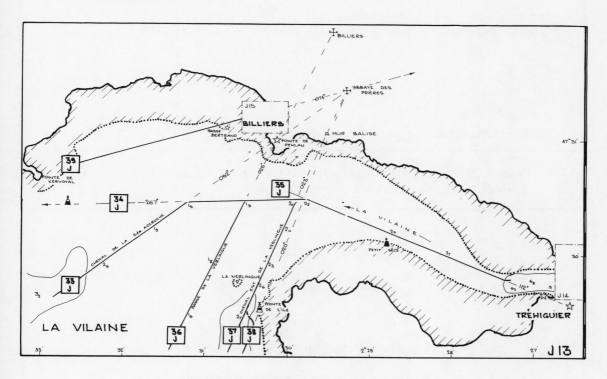

1—Chenal de la Grande Accroche (1·5m)

On the north shore of the estuary is a mark used often, the tall square Tour de l'Abbaye des Prières. We now use it for

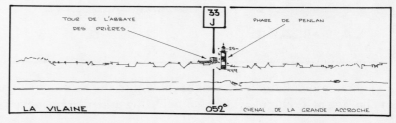

33J the Abbey × Penlan light. Then take

47

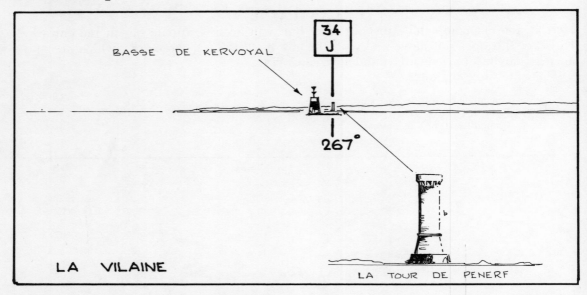

34J La Tour de Penerf (a disused lighthouse) to the right of Basse de Kervoyal *tourelle*. The final mark which takes you into much deeper water is

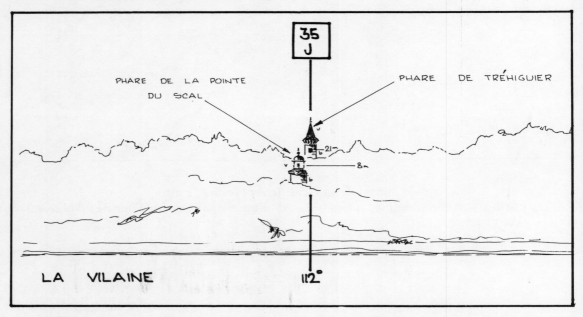

35J Tréhiguier light to the right of Pointe du Scal light. This takes you almost to the anchorage at Tréhiguier, the channel being adequately buoyed.

2—Passe de la Varlingue (1·6m)

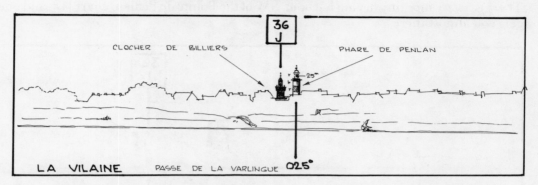

36J Billiers church × Penlan lighthouse which joins 34J

3—Chenal Est de la Varlingue (2·0m)

This is the deepest of the three channels but the front mark, in a bad light, needs a microscope:

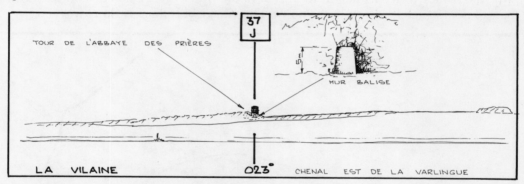

37J the Abbey × a small white *amer*, which likewise joins 34J. A useful safety mark to keep you off the several dangers of the coast to the southward is

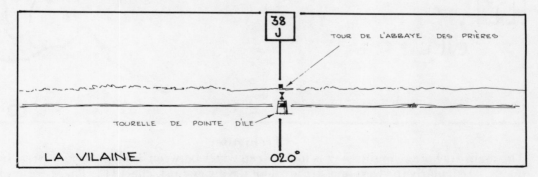

38J the Abbey × the Pointe de l'Ile *tourelle*. Don't open it to the right.

Billiers

There is a very nice tiny drying harbour NW of the Pointe de Penlan, chart J15, and one mark gets you almost there

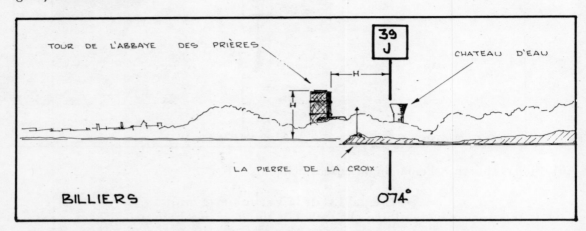

39J the Abbey tower its height to the left of the water tower. You'll soon see a row of stb'd hand *balises* which lead you to the quay. The village of Billiers is 2km away.

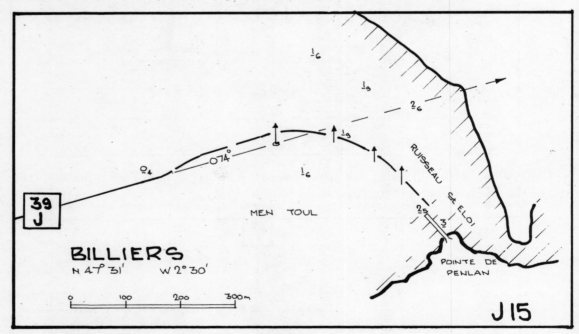

Tréhiguier

The main anchorage in the river is in the deep water between Tréhiguier and Arzal but the quays, particularly the west one, are ideal for drying against, chart J14. There are few shops in Tréhiguier but the next big town up the river is La Roche Bernard.

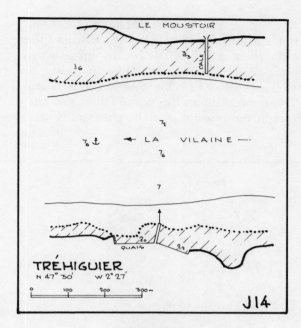

Port d'Arzal

A recent development, the barrage and lock have made the upper reaches of the river Vilaine into one vast boating lake between 3·0m and 5·0m deep. The lock opens on the hour, except 13.00, every day between 07.00 and 20.00. Chart J16 gives all details and there are convenient alongside waiting places upstream and downstream in addition to many buoys. Some 50 boats are moored on the left bank upstream near the barrage.

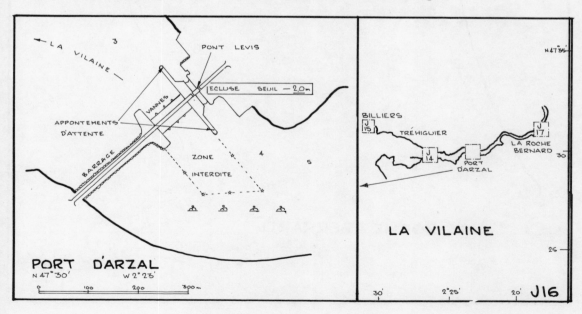

French Pilot 4

La Roche Bernard

Further up the river is La Roche Bernard, chart J17, where there are two marinas and unlimited anchoring in the river. The Quai de St Antoine is usually very crowded as is the Nouveau Port, but if you want an alongside drying berth, try the river quay, SE of La Truie. From La Roche Bernard it is possible to follow the river to Redon and to lock in to the Canal d'Ille-et-Rance which takes you through Rennes to Dinan and the river Rance for St Malo. The canal limits are: draught 1·2m, beam 4·5m, height 2·5m. There is no commercial traffic.

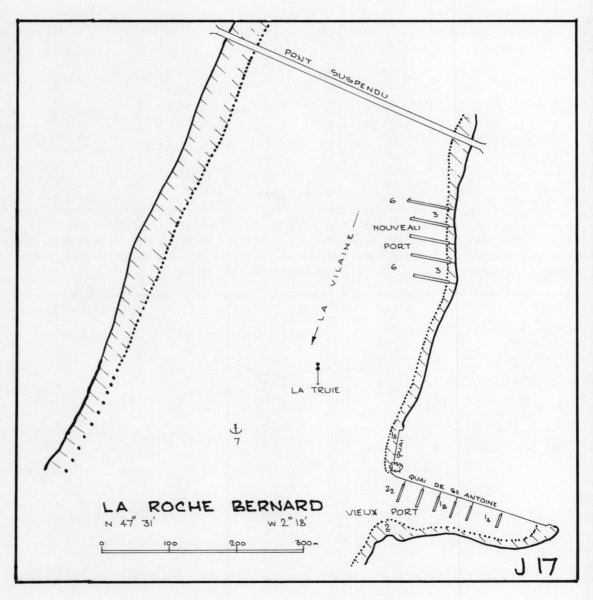

LA ROCHE BERNARD

N 47° 31'   W 2° 18'

0   100   200   300 m

LE MESQUER

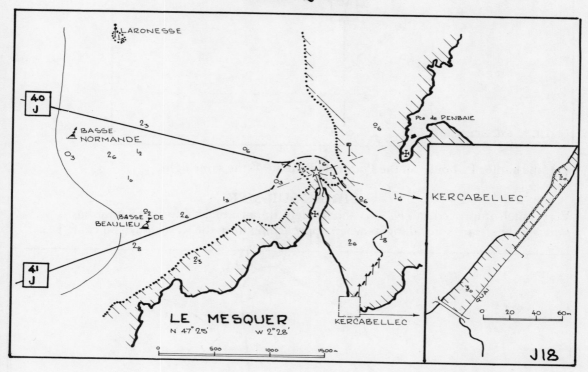

A small harbour among the oyster beds and marshes, with two entrances, chart J 18. The first is

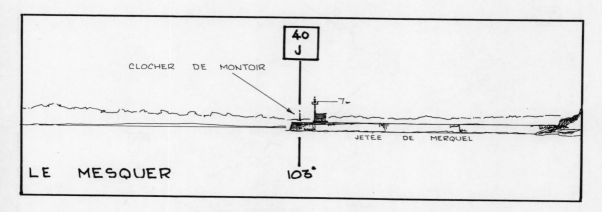

40J Montoir church × the light at the end of the jetty. When 2 cables off the light, make a handrail north, and take the track shown. It is marked by *balises* and withies for another mile up to the tiny port of Kercabellec. An alternative entrance south of the Basse de Beaulieu is

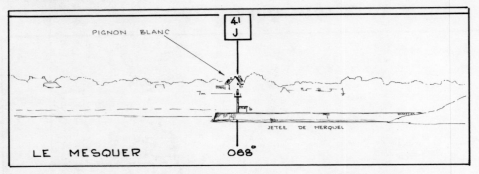

PIGNON BLANC

LE MESQUER   068°

JETÉE DE MERQUEL

41J the gable of a house on the Pointe de Penbaie ✕ the same light.

PIRIAC-SUR-MER

Very much improved lately, this moderately important fishing town now has a *port de plaisance*. There are some dangers when approaching from the west shown on chart J19.

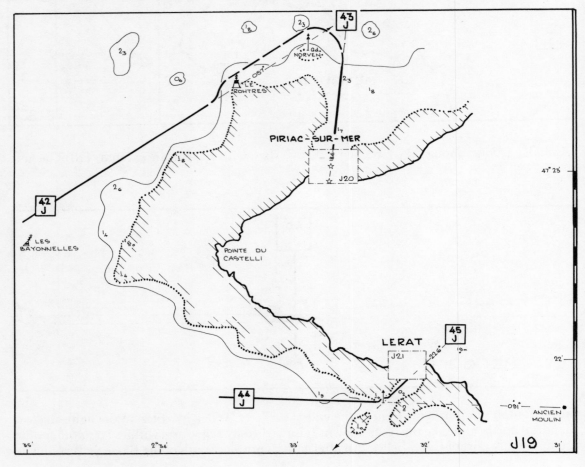

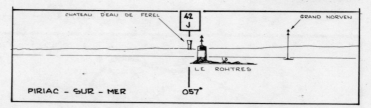

42J Ferel water tower to the left of Le Rohtres *tourelle* (CN). When 2 cables off, follow the track shown leaving Norven to stb'd until

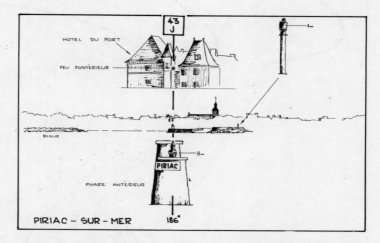

43J the two lights in line. The back light is coyly hidden between two buildings to the left of the church. The fishing activity is concentrated at the north quay and on the quay to the south of the light, but the west end near the *3m* sounding is usually clear. The *port de plaisance* is SE of the new digue, chart J20.

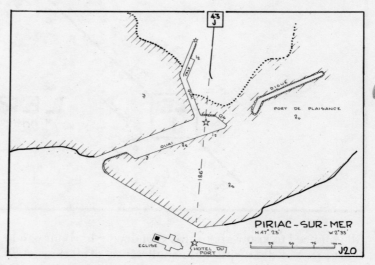

LERAT

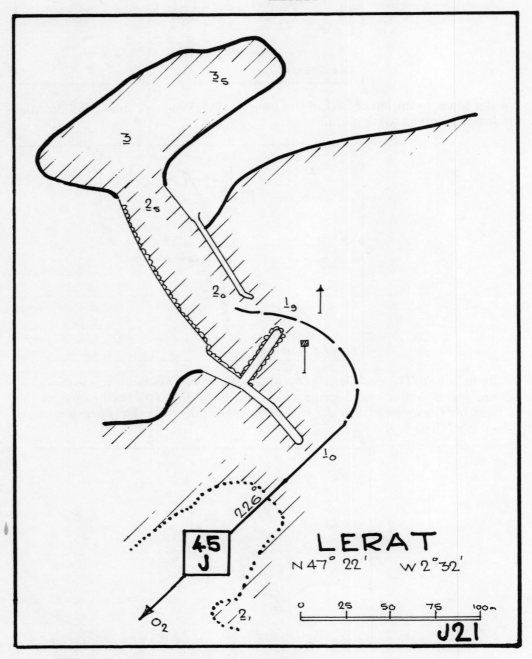

LERAT

N 47° 22'   W 2°32'

0   25   50   75   100m

J21

All you need to know about this vast drying harbour can be gained from the scale shown on chart J21. You did mention the soullessness of marinas, didn't you? Come in from the west on

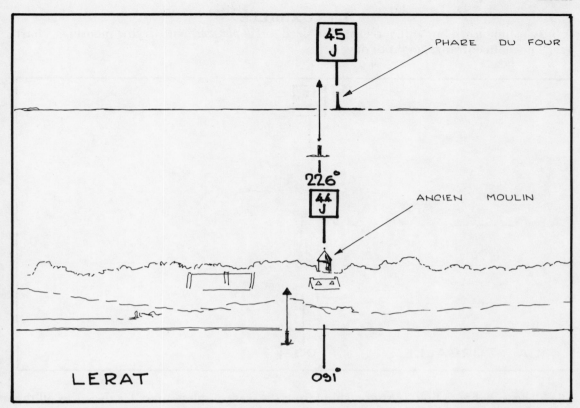

44J an old mill to the right of the stb'd *balise* chart J19 (p.54). Dodge to the north of it and look for a back mark (on the same sketch)

45J the Four lighthouse ✕ the same *balise*.

LA TURBALLE

A busy fishing town but with a reasonable sized *port de plaisance* with drying moorings, chart J22. The main entrance, night or day, is

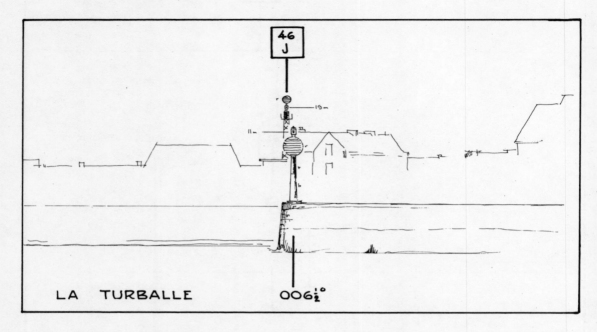

46J the two lights in line. If approaching from the west, a safety mark for the rocks off the coast is

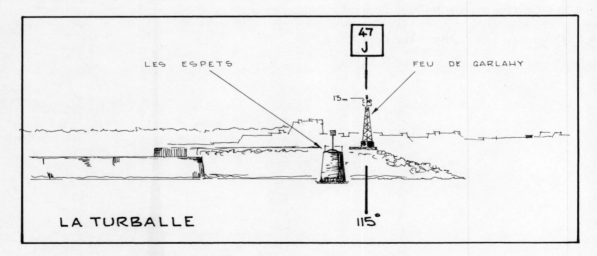

47J the Garlahy light structure seen outside Les Espets *tourelle*.

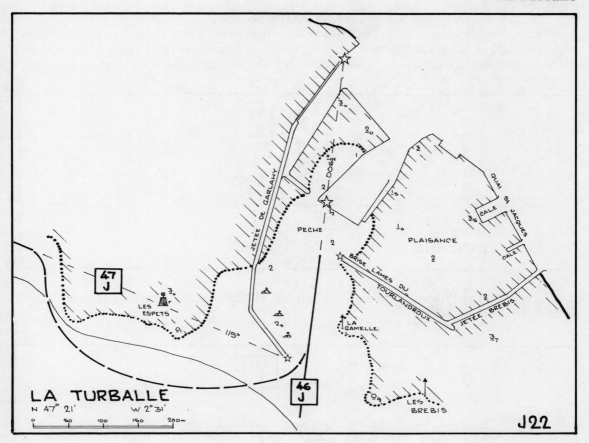

LA TURBALLE

N 47° 21'　　W 2° 31'

0　50　100　150　200m

J22

LE CROISIC

A fishing harbour of diminishing importance, its quays were built of ships' ballast in the days of the flourishing salt trade from the adjacent marshes. Shown on chart J23, it is the port for the beautiful walled town of Guérande 8km away. Coming from the west, a safety mark is

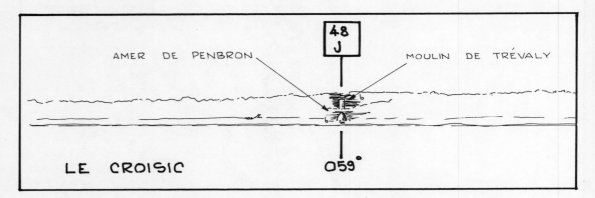

48J the old mill at Trévaly × the *amer* of Penbron. The first mark for the long channel, which is dredged to 1·2m, is

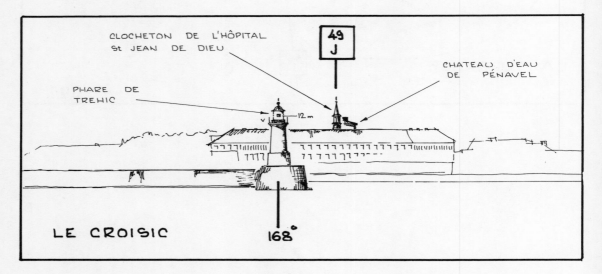

49J the *clocheton* de L'Hôpital St Jean de Dieu × Le Tréhic light. Penavel water tower is almost on the same transit. When 3 cables from the lighthouse, take the first of three highly decorative transits as good at night as by day

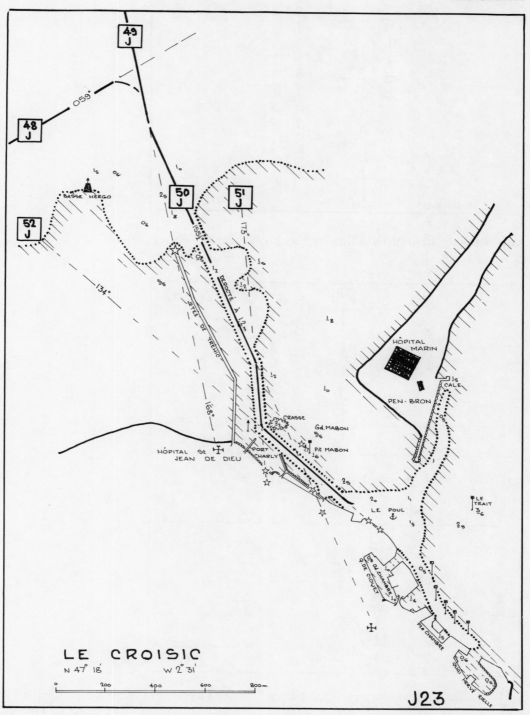

LE CROISIC

N 47° 18'   W 2° 31'

0   200   400   600   800m

J23

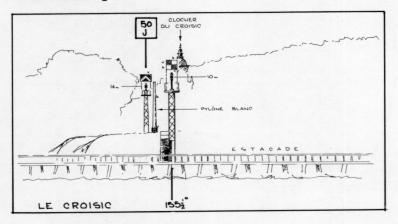

50J two light structures in line with a church almost behind them.

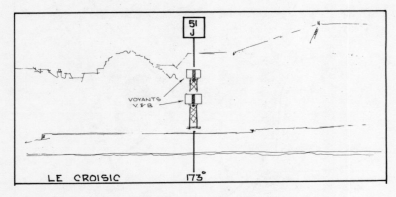

51J two white panels with green stripes in line

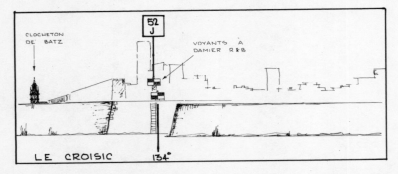

52J two red and white chequered panels in line. This takes you to Le Poul where you can anchor or take a buoy. The *port de plaisance* is in the last basin of all at the Quai Hervé Rielle.

BATZ-SUR-MER

This not particularly sheltered harbour houses a few boats but is a convenient summertime drying stop before coming upon its fashionable neighbours round the corner, chart J24. The western approach is

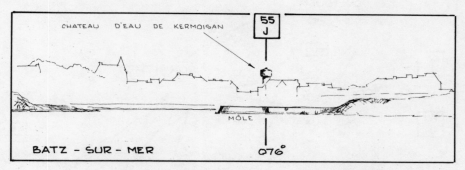

55J Kermoisan water tower × the bend in the middle of the mole. The alternative is

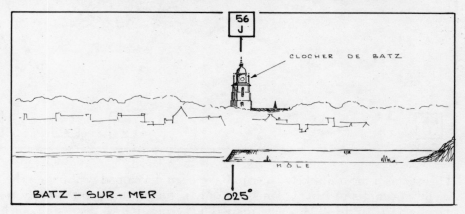

56J Batz church × end of the mole. The east end of the harbour is clean and sandy.

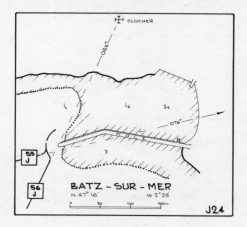

LE POULIGUEN

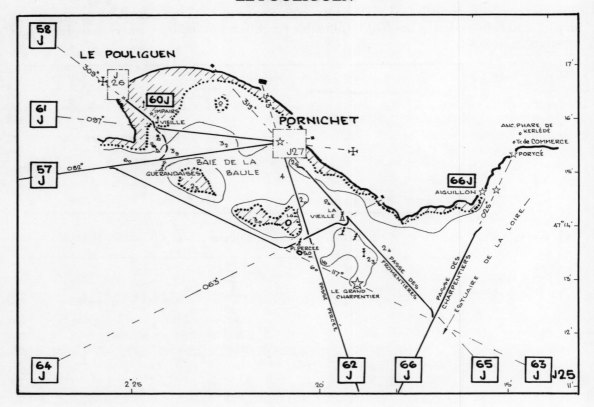

This seems a most unlikely river to have been developed well before the war as one of the yachting centres in Biscay, since not only is the entrance not entirely easy, but the whole narrow harbour dries. However there is no accounting for tastes as 3,000 boat owners show. Coming from the west, first find yourself between the two lit buoys of Les Guérandaises and Basse Martineau (white sector of the Pornichet light), chart J25. You should then be on

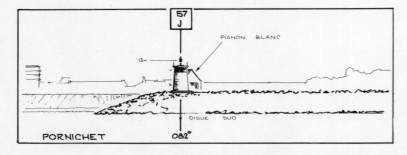

57J a white gable just to the right of Pornichet light. The channel, first north then NW, is buoyed and

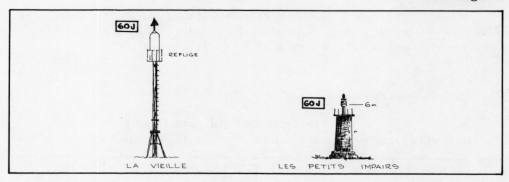

60J shows views of the two stb'd marks. Having left La Vieille *balise* and refuge 50m to stb'd take

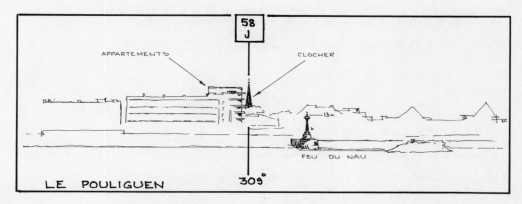

58J the church just to the right of a prominent block of flats. You are now on the drying part and if you want to gain every millimetre enter the harbour, chart J26, (p.67), on

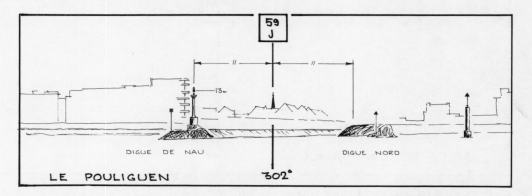

59J the church spire in the middle of the harbour entrance. The right bank is less commercialised and much more convenient for shopping than the La Baule side. Berths can often be found.

PORNICHET

This 1,500-place marina must have done a lot to reduce the pressure on Le Pouliguen. There are three approaches all shown on chart J25 (p.64):

> 1—from Le Pouliguen
> 2—from Passe Percée
> 3—Passe des Fromentières

1—from Le Pouliguen (1·3m)

This is a slightly needless mark which takes you straight across the bay

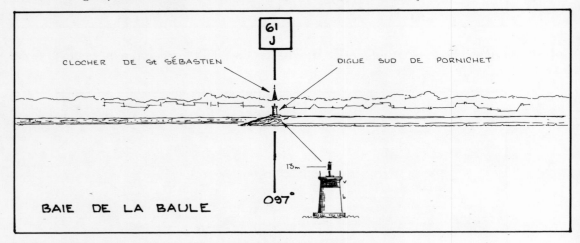

61J St Sébastien church × the south breakwater light.

2—from Passe Percée (1·5m)

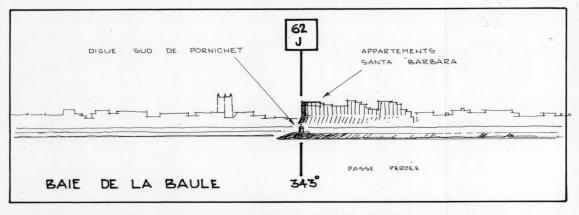

62J The left hand side of a prominent multiblock of apartments × the south pierhead light.

3—Passe des Fromentières

From the west, first take the safety mark for all the rocks in the Baie de la Baule,

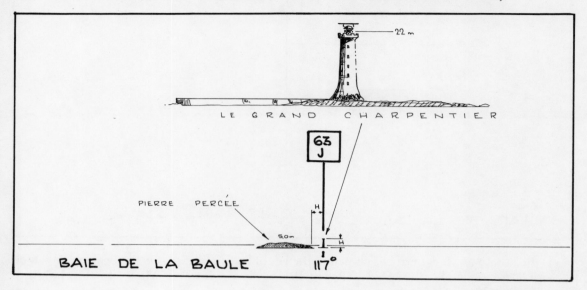

63J Grand Charpentier lighthouse to the right of La Pierre Percée. When 3 cables short of this rock, turn to port on the Passage du Ronfle,

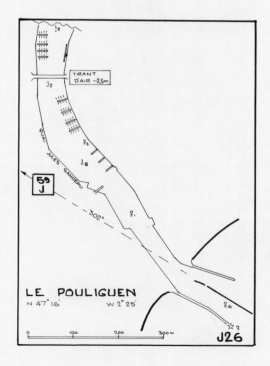

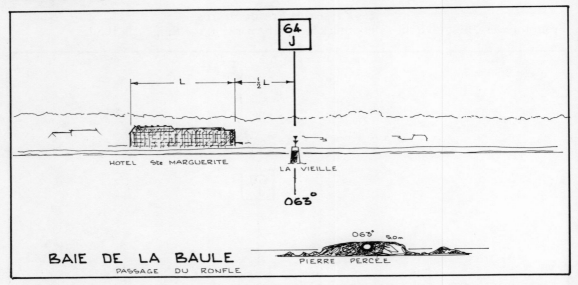

64J the hotel Ste Marguerite as shown to the left of La Vieille. This is the bearing on which you will see the hole through La Pierre Percée from seaward. Make a cable handrail east-about of La Vieille and take

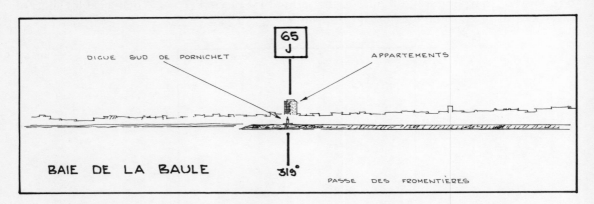

65J a high rise building in La Baule × the Pornichet south pierhead light. The marina is shown on chart J27. There is of course every conceivable facility.

Cowering behind all this glitter is the old harbour of Pornichet, also shown on chart J27, and from this you will note the well-marked channel. The east side of the quai is best for drying out against; mud and stones.

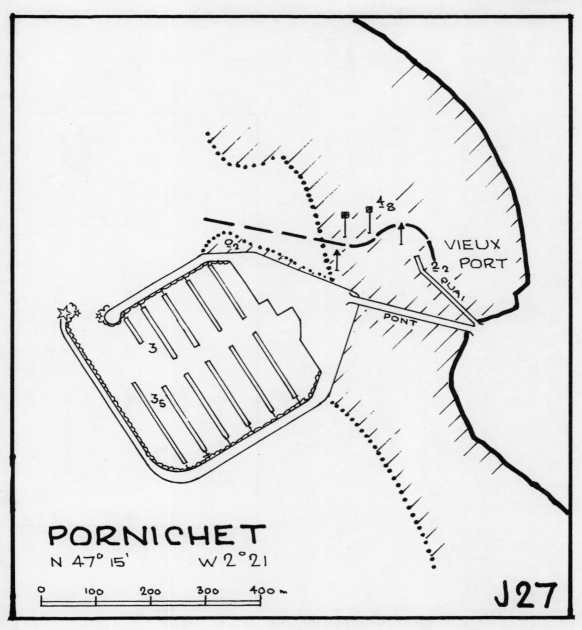

PORNICHET

N 47° 15'   W 2° 21

0   100   200   300   400 m

J27

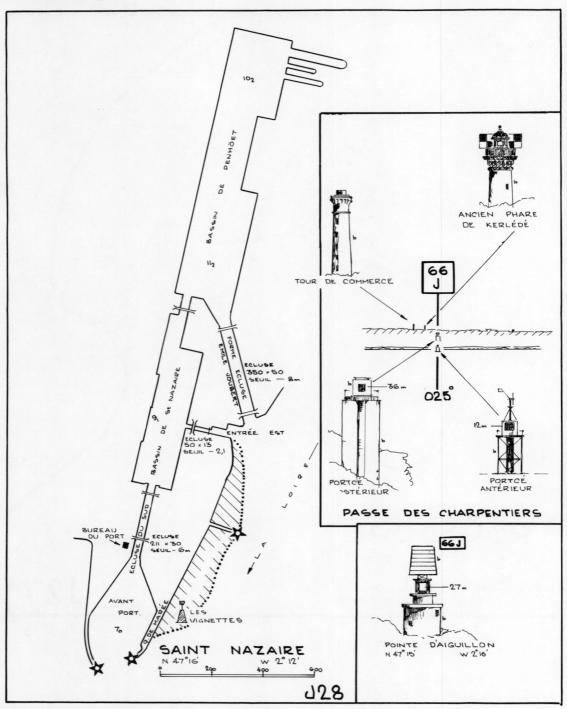

SAINT NAZAIRE
N 47°16'   W 2°12'

J28

PASSE DES CHARPENTIERS

POINTE D'AIGUILLON
N 47°15'   W 2°16'

ST NAZAIRE

Yachtsmen are tolerated but only just in this enormous commercial harbour which builds some of the largest ships in France. It is shown on chart J28. The entrance mark is the Passe des Charpentiers

66J also shown on the above chart. The two Portcé lights, lit during daylight, in line. This is the main transit for the Loire river shown on chart J25 (p.64). The east entrance must be used. The lock is opened on demand, day or night. Apply to the *Bureau du Port*.

The rest of the river Loire is treated in the next section.

K   La Loire to Jard-sur-Mer

This is a section of contrasts, the enormous river of the Loire, the vast bay of Bourgneuf with its oysters and mussels, the rugged Ile d'Yeu and the inhospitable coast of Pays de Monts.

We will start off with the Loire, that slow-moving shallow waterway, ascending the right bank and descending the left. Chart K1 shows all the riverside ports which are large enough to have some sort of a quay. None of them require special comment and all have an old-world charm dating from the days before road transport.

COUËRON	Chart K2
BASSE INDRE	Chart K3
HAUT INDRE	Chart K4
NANTES	Chart K5
LE PELLERIN	Chart K6
PAIMBOEUF	Chart K7

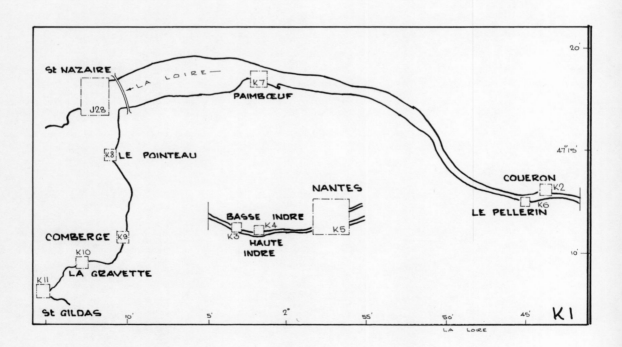

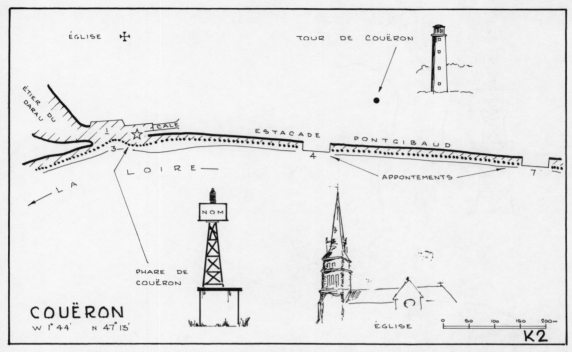

ÉGLISE

TOUR DE COUËRON

ÉTIER DU DARAU

CALE

ESTACADE   PONTGIBAUD

LA   LOIRE

APPONTEMENTS

LA

PHARE DE COUËRON

NOM

ÉGLISE

COUËRON
W 1°44'   N 47°15'

0   50   100   150   200m

K2

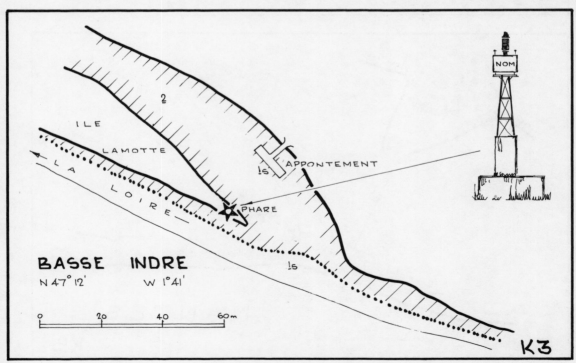

ILE

LAMOTTE

LA   LOIRE

APPONTEMENT

NOM

PHARE

BASSE INDRE
N 47°12'   W 1°41'

0   20   40   60m

K3

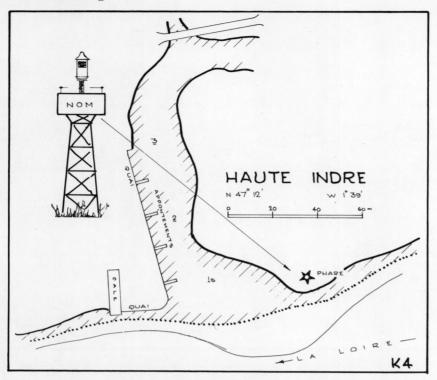

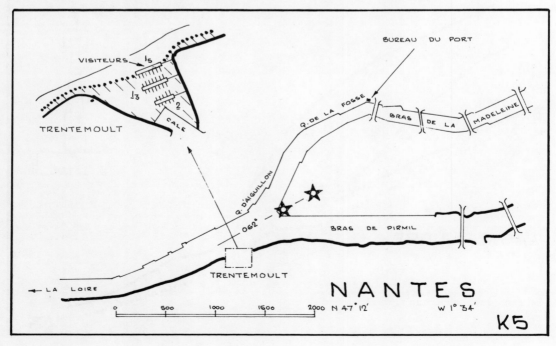

The chart of Nantes shows the oddly small *port de plaisance* of Trentemoult which not only is some way out of the town but dries completely. It is possible you can find a berth in the town either on the Quai d'Aiguillon or on the Quai de la Fosse by going to the *Bureau de Port*.

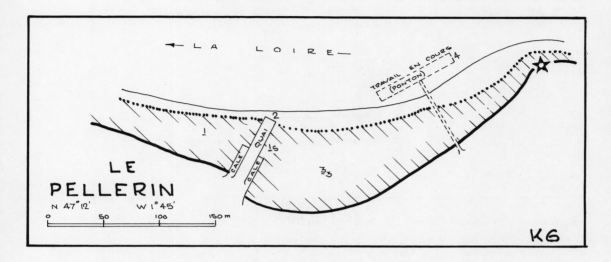

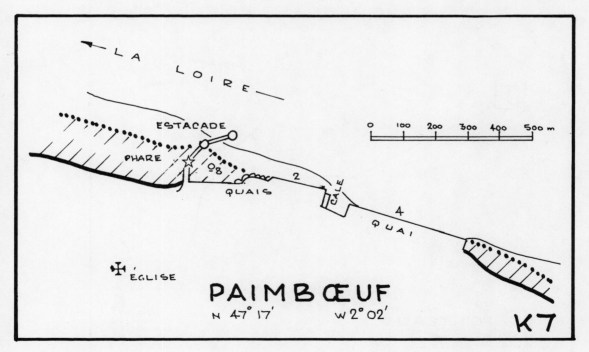

75

French Pilot 4

LE POINTEAU

This is a new development which I am assured goes by the name of Le Pointeau rather than the town it serves, St Brevin l'Océan, chart K8. The whole area dries and the only mark is

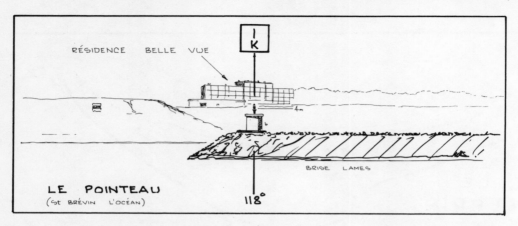

1K the centre of some flats × the pierhead light.

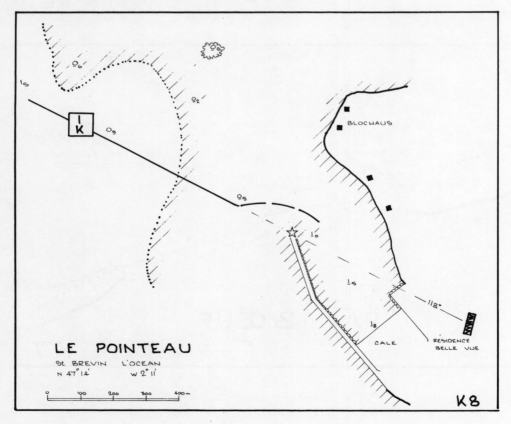

COMBERGE

Another dying fishing harbour brought up-to-date, chart K9. There are rocks near the approach so use

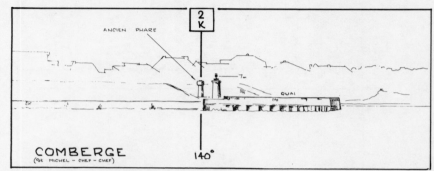

2K the old lighthouse to the left of the new. This is the harbour for the nearby town of St Michel-Chef-Chef.

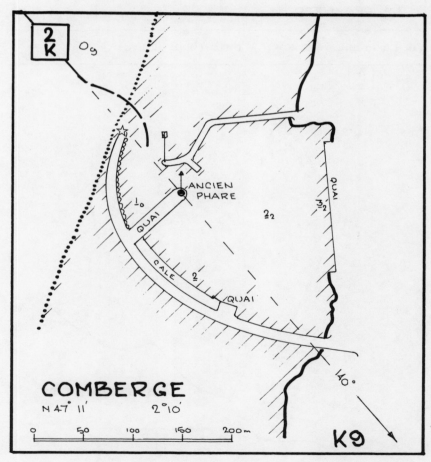

LA GRAVETTE

Chart K10 shows this new harbour largely used by fishing boats. The entrance mark is

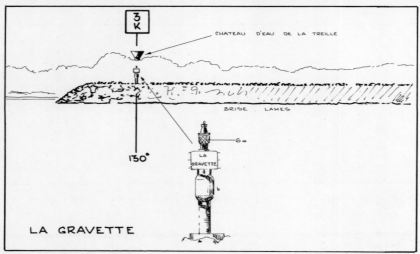

3K La Treille water tower × the harbour light.

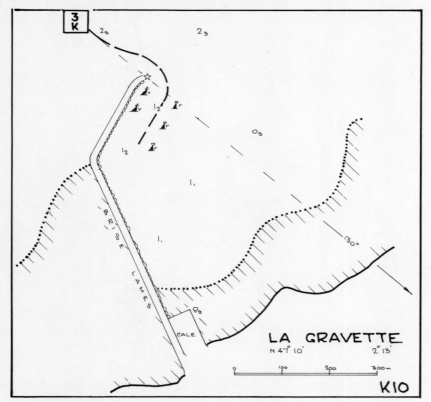

ST GILDAS (Loire)

A most useful refuge stuck out into the Loire estuary, which unfortunately has no nearby village, chart K11. The entrance is the same as the white sector of the light and is

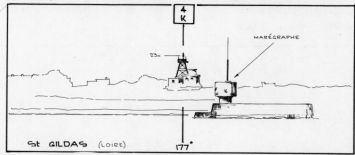

4K the lighthouse structure to the left of the *marégraphe*. This machine existed long before the harbour and repeats, minute by minute, the tidal heights by radio to the harbour office at St Nazaire.

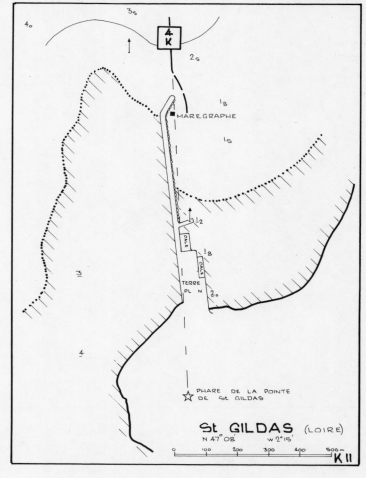

French Pilot 4

PORNIC

We are now in the ten-mile-wide Baie de Bourgneuf, the east and south sides of which are full of rocks, mud banks or, worse, *bouchots de moules*. These are lines of wooden posts used for the cultivation of mussels and will be seen from here right down to the Gironde. The whole bay is shown on chart K12.

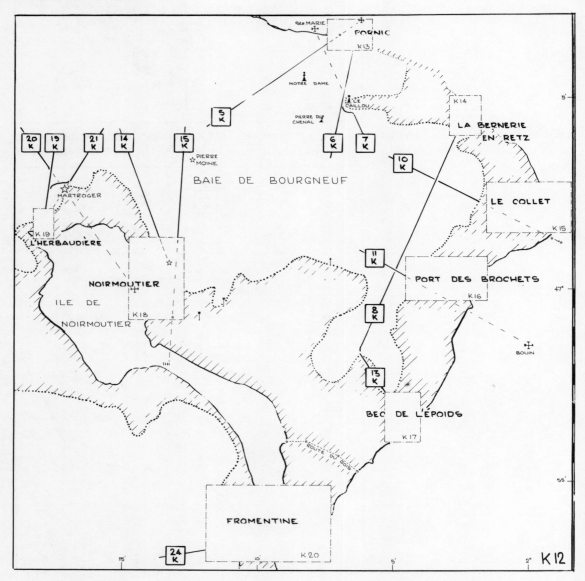

The first port we come to, Pornic, is a most interesting old town with a large marina for 1,000 boats at the entrance, the first of many now built down this coast, chart K13. The main entrance is from the SW

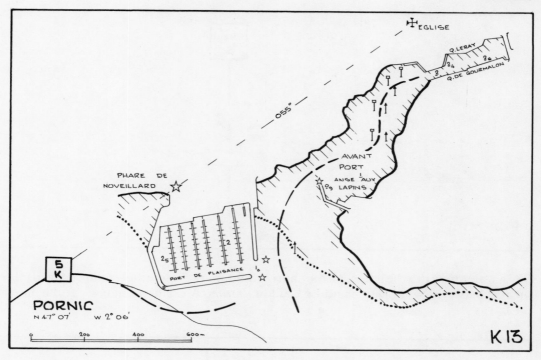

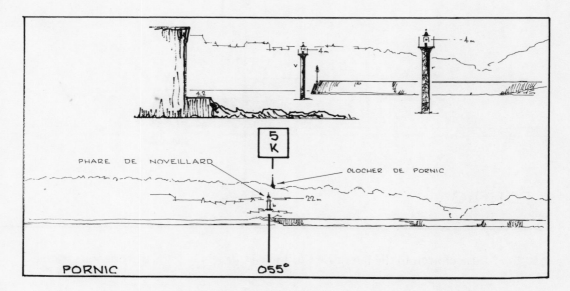

5K the church × Noveillard lighthouse. On this sketch is shown the end of the south pierhead of the *port de plaisance*. Note the two light structures which form a gateway to the marina. An approach to Pornic from the middle of the bay is

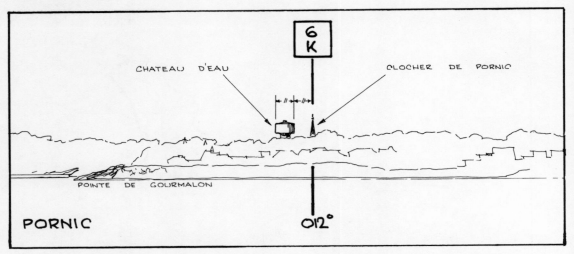

6K a water tower to the left of Pornic church. This mark if accurately followed leads through the reef between Notre Dame and Le Caillou *tourelles*. A useful mark to clear the dangers in the bay is

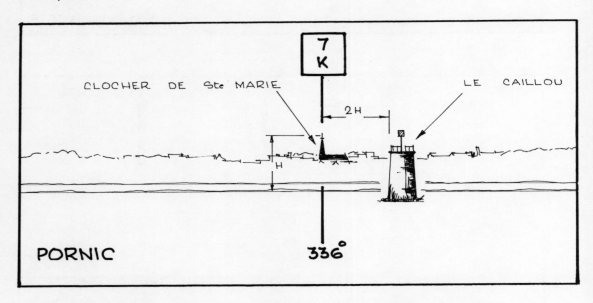

7K Ste Marie church to the left of Le Caillou *tourelle* which joins the previous transit.

LA BERNERIE-EN-RETZ

A pleasant little drying harbour with a shallow entrance, chart K14. From far off, the approach is

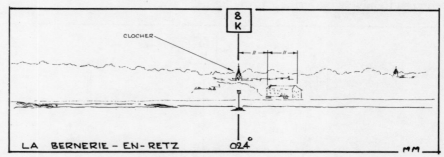

8K the church to the left of a large warehouse. A port hand *balise* is on the same transit, and when within a couple of cables off this take

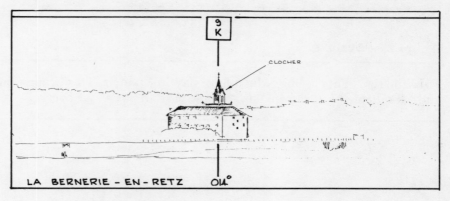

9K the church × the warehouse.

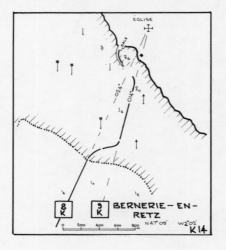

LE COLLET

Here is the first of three tiny shellfish ports which can be entered mostly above half tide; the marks given take you through those forests of *bouchots*, chart K15. This one is important enough to have wonderful leading lights,

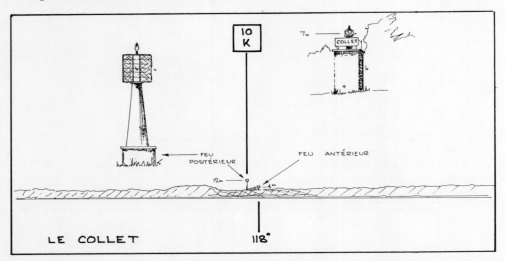

10K the two lights in line, which have daytime panels. This brings you between Baranger and Mouillepied *balises* from where a line of stb'd posts take over.

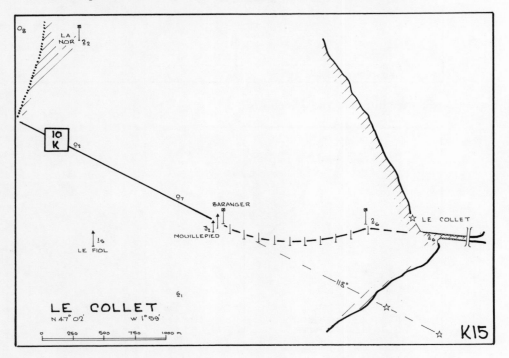

PORT DES BROCHETS

This has a dog-leg approach, chart K16, the first mark being

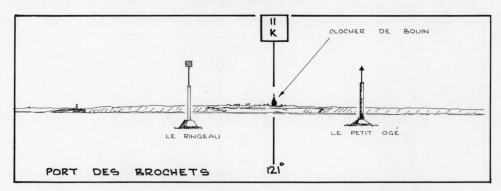

11K Bouin church in the entrance gateway of Le Ringeau and Le Petit Ogé *balises*. When through this gateway, turn to port on

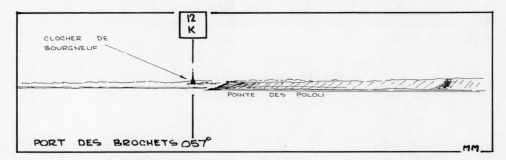

12K Bourgneuf church open of Pointe des Polou. Turn when looking down between the line of posts (at night the 3° white light sector)

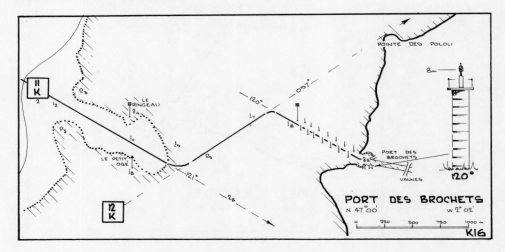

85

BEC DE L'ÉPOIDS

Originally the port for Beauvoir-sur-Mer, this is now devoted solely to oysters, chart K17. The approach is the same as the 1° sectored light

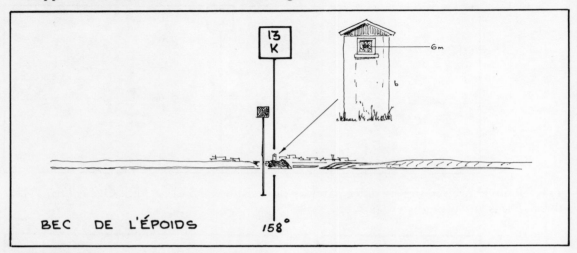

13K the lighthouse × the port hand *balise*.

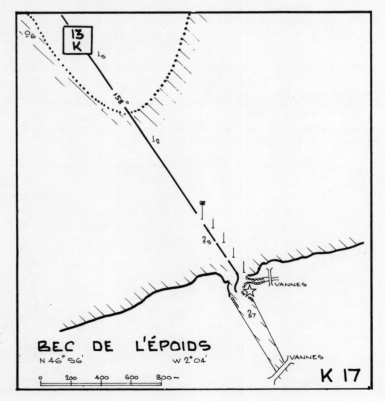

86

NOIRMOUTIER

It's now time we went across to the Ile de Noirmoutier, traditionally 'as much water as land'. Connected for two hundred years by a causeway, route du Grois, it has now been spoilt, so the inhabitants say, by an enormous motorway bridge over the Goulet de Fromentine. There is a very secure anchorage at Bois de la Chaise, shown on chart K18. The wooden wharf is the berth for the ferries from Pornic but many mooring buoys have been laid in deep water. The approach from the north is through reefs so use

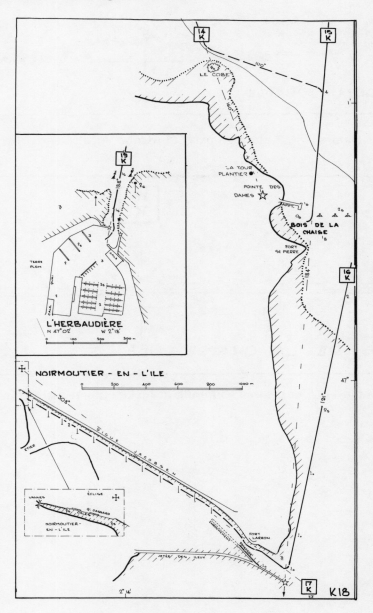

87

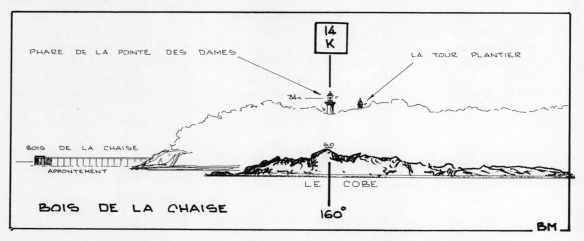

14K Pointe des Dames light × the high part of Le Cobe. An alternative from further out in the bay is

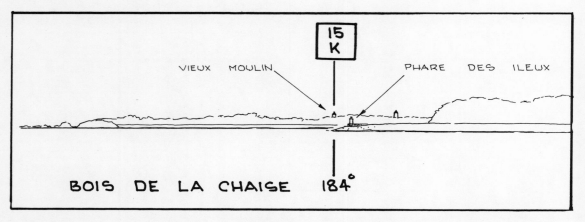

15K the easternmost of four disused windmills at La Guérinière just to the left of the Ileux light.

Noirmoutier-en-l'Ile

The port of Noirmoutier, formally a flourishing fishing harbour, is a pleasant little town up a short canal, chart K18 (p.87). The approach can be made from Bois de la Chaise on

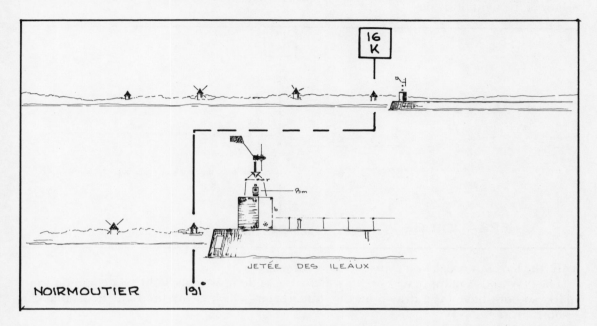

16K the westernmost of those same four windmills just to the left of the Jetée des Ileux. To avoid the sandbank don't turn too quickly into the canal, but take

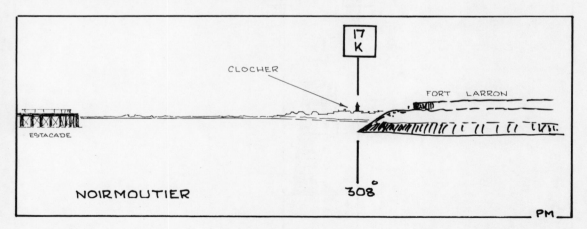

17K Noirmoutier church open to the left of Fort Larron. You are now in the canal, the south side of which is marked by a row of posts. Keep in the middle; the berths are right in the centre of the town on the Quai Gassard.

French Pilot 4

L'Herbaudière

This is an old fishing village turned marina which has had a recent facelift, chart K18 (p.87). Back to chart K12 (p.80) for the three approaches. The main one is in the white sector of the light or by day

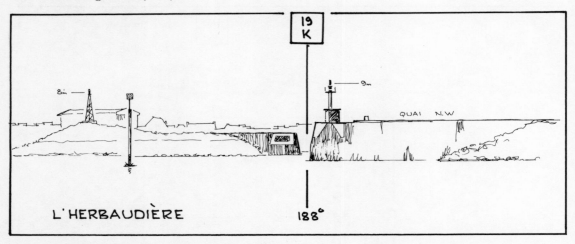

19K the harbour breakwaters just open as shown.

The NW and NE approaches both use the Basse de Martroger light as a common front mark, so both have been drawn on the same sketch which clears the Banc de la Blanche (1·8m). From the NW

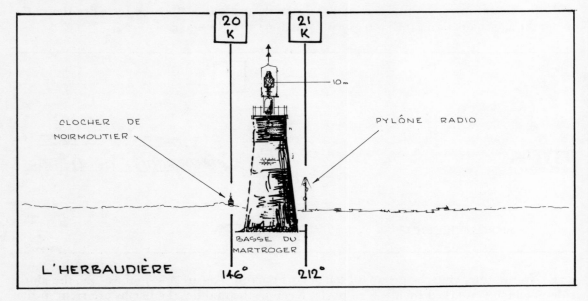

20K the church × the left of the light; and from the NE
21K a telecommunications antenna to the right of the light.

FROMENTINE

Because of the causeway, route du Grois, this is only accessible from the west under the new road bridge. The ferry from here to Ile d'Yeu, whose terminus this is, enters only 1½ hours before and after high water, St Nazaire. There is a bar extending a mile and a half west of the bridge but the ferry skippers assure me that the position of the channel moves extremely seldom, certainly no more often than 7–10 years. If it does move, one hopes the buoys will be repositioned. The best time to enter for a stranger is about 1 hour before high water, St Nazaire, and the channel is on chart K19.

Starting off from the lit landfall buoy, look for the only transit,

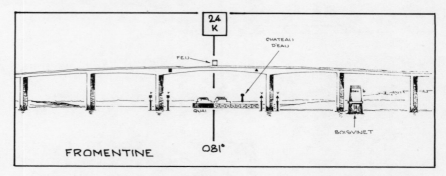

24K the NW corner of the landing stage exactly central with the middle arch of the bridge which leaves buoys '1' and '3' to stb'd. When 2 cables from '5' buoy, steer between it and '4' roughly NE. Leave the lit Milieu *tourelle* 50m to port and steer for the centre bridge arch. By now you are well over the bar, and as soon as you are under the bridge you can anchor or take one of the many mooring buoys. The landing stage is for the exclusive use of the Ile d'Yeu ferries. There is, however, on the north side a disused slip, a relic of the days before the bridge was built. Only well-insured and powerful swimmers should try this entrance at night, or with much onshore wind.

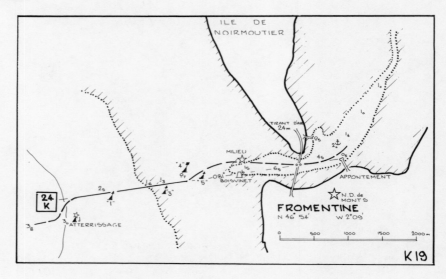

ILE D'YEU
Port Joinville (Port Breton)

Site of a settlement by the Druids, this island is more akin to the cliffs of north Brittany from where incidentally it was colonised in the middle ages. The chief town and port is Joinville, chart K20. The harbour has recently been extended considerably and also has a new marina. The pair of leading lights have been replaced by a new directional light on the Quai Canada but our main daytime mark will be the old transit

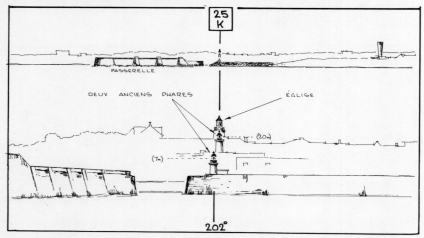

25K the church and the two old lighthouses in line. There are some rocks near the entrance so if approaching from the west use as a safety mark

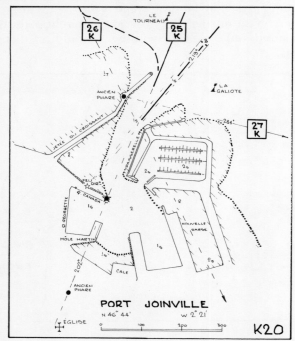

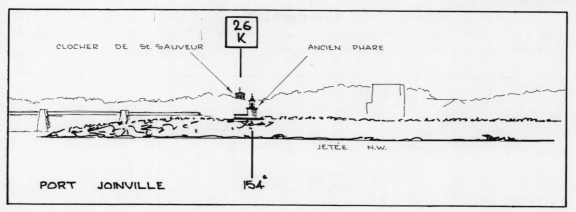

26K St Sauveur church to the left of, but never touching, the old lighthouse on the north quay. If coming from the east a safety mark for the rocks off the coast is

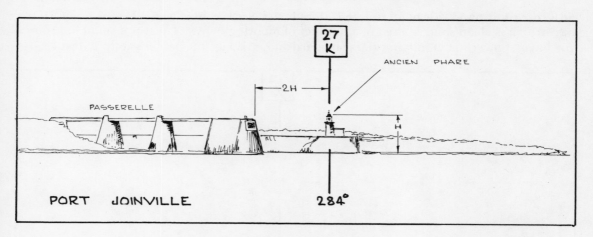

27K the same old lighthouse well to the right of the *passerelle* as shown. Apart from the marina it is difficult to find an area not used by fishing boats.

Anse des Vieilles

This is the first of two charming little harbours on the south coast shown on chart K21. To clear the rocks on either side of the entrance, use

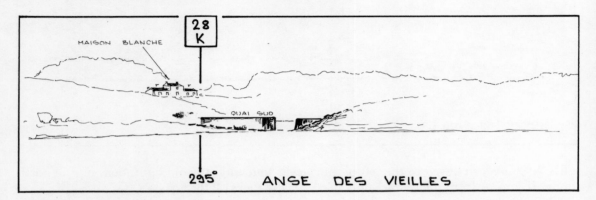

28K the right-hand side of an isolated white house ✕ the root of a small stone quay. In the bay, which is clean sand, there are a number of mooring buoys. There is a small harbour if you have a powerful magnifying glass on board and, to dodge the pebbles at the entrance, use

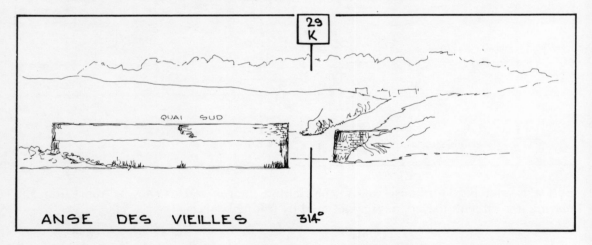

29K a pile of stones to the left of the roadway in the middle of the harbour mouth. Have you ever been in a smaller harbour?

La Meule

I have left this gem on Ile d'Yeu to the end. It is a small very enclosed creek on chart K21 which, unlike its counterpart in Belle Ile, Stêr Ouen, has a proper quay. To lead you in among the rocks take

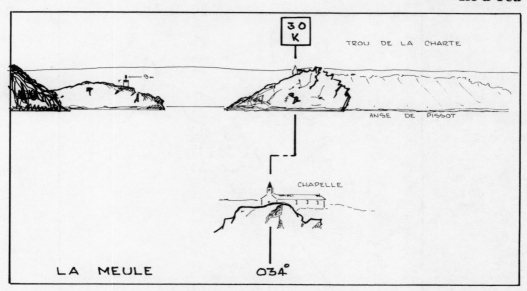

LA MEULE   034°

30K an isolated chapel ✕ the top of a high rock Trou de la Charte. When 100m off this rock, the harbour entrance can be seen, chart K22, (p.96). A number of fishing boats have summer moorings outside near the 4m sounding but, because it dries early on, the north end of the harbour is often clear. There are no supplies nearer than St Sauveur, 2km away.

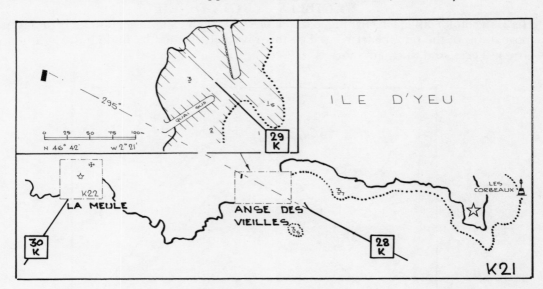

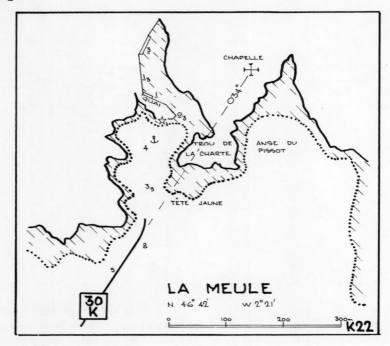

LA MEULE

N 46° 42' W 2° 21'

ST GILLES—CROIX-DE-VIE

Here is a comfortable, part fishing, part *plaisance*, harbour, easy to get into night or day and at most states of the tide, chart K23. The town is easily located by finding the lighthouse on Grosse Terre, and the main mark is

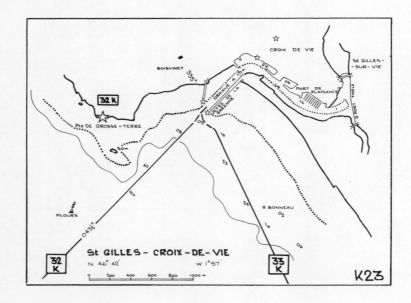

St GILLES - CROIX-DE-VIE

N 46° 42' W 1° 57'

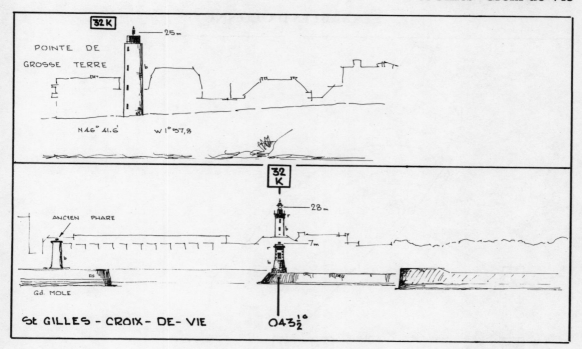

32K the two harbour lighthouses in line. To keep you clear of the Rocher Bonneau, or when approaching from the SE use

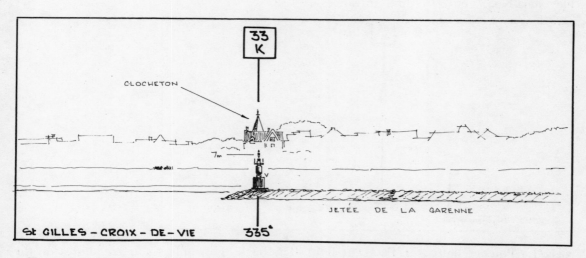

33K a villa with a steep spire × the light on the Jetée de la Garenne. A considerable fleet of trawlers is based here but if the marina is full a drying berth may be found on the Quai Port Fidèle on the St Gilles side of the river.

LES SABLES D'OLONNE

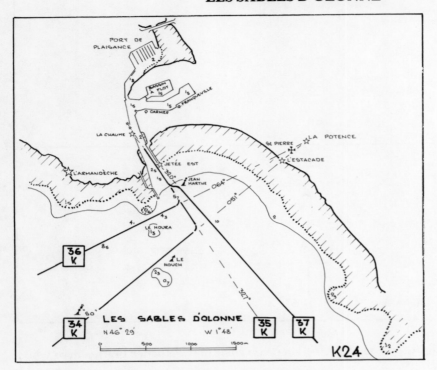

Thanks to a large new marina, Port Alona, this former fishing port has taken a new lease of life, chart K24. There are three entrances

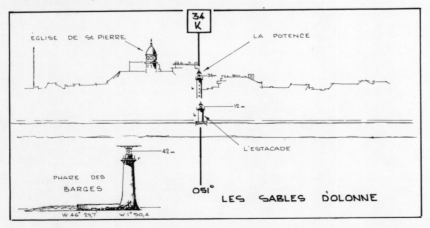

34K La Potence light × L'Estacade light. The back light, unless you are almost on the transit, is obscured by high buildings. These lights are on all day. To locate the town on the rather featureless coast, I show on the sketch a view of Les Barges lighthouse. As soon as the harbour entrance opens up, take

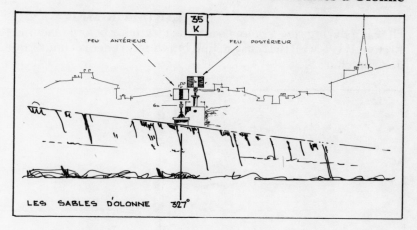

35K a pair of lights in line with red and white topmarks which takes you safely into the dredged channel. An alternative mark leading north of Le Noura, by day only, is

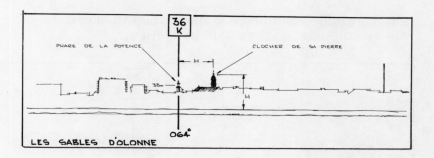

36K the back light, La Potence, open to the left of St Pierre church. Coming up from the south, take

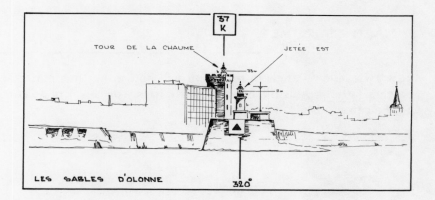

37K Tour de la Chaume, with a light built on top, × the east jetty head.

French Pilot 4

JARD-SUR-MER

Here is a drying *port de plaisance,* the last harbour before reaching the shelter of the Pertuis de Breton. It is not all that easy to find but the entrance is about 5 miles NW of Pointe du Grouin du Cou, chart K25.

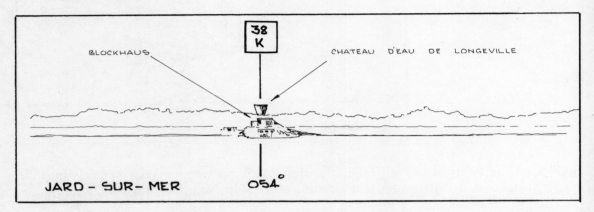

38K Longeville water tower ✕ an old German blockhouse on the beach. When about 4 cables off the coast, make an acute port turn to

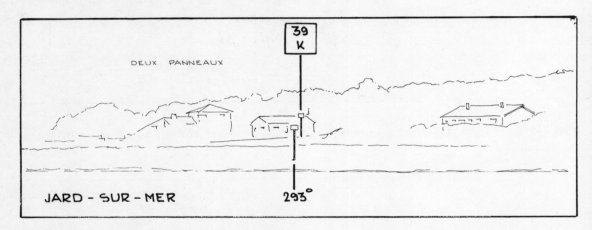

39K two yellow panels in line. You will notice that this mark takes you over two drying patches (*0·1m*) but as the harbour itself dries *2·5m* this is of no consequence. The bottom is hard clean sand.

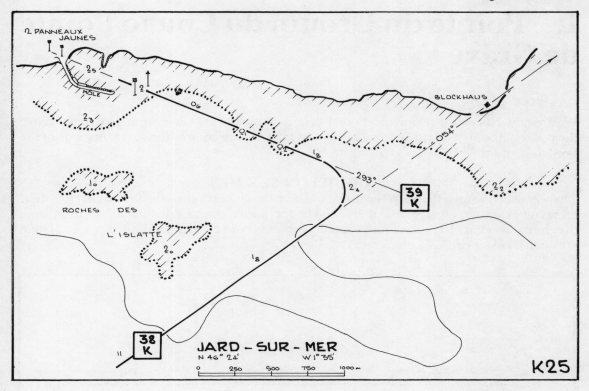

2 PANNEAUX JAUNES

BLOCKHAUS

MÔLE

2_5

2_2

06

0_4

1_8

293°

2_4

054°

2_2

39 K

2_3

ROCHES DES

1_0

L'ISLATTE

2_0

1_8

38 K

11

JARD – SUR – MER

N 46° 24'   W 1° 35'

0   250   500   750   1000 m

K25

L Pointe du Grouin du Cou to Pointe de Grave

We are now entering a region of a very different character: les Charentes. The coast is low; the houses take on a Spanish appearance and our journey ends in the vast vineyards of the Bordelais and the Medoc.

L'AIGUILLON-SUR-MER

The whole of this area between the mainland and Ile de Ré is called the Pertuis Breton. In it life revolves around bivalves and there is no better place to appreciate this vast industry than the river Lay, chart L1. First find yourself 3 cables NW of the landfall buoy, Le Lay. There are no marks so I can only give you

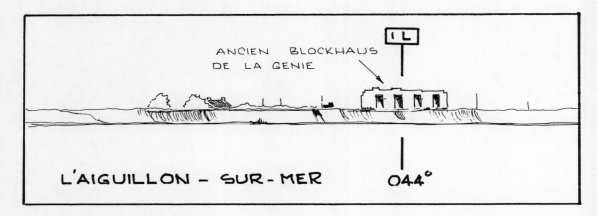

1L an old German blockhouse on the seawall at 044°. From here on you should be able to pick out the channel which is marked both sides by withies and an occasional buoy. Five miles from the landfall buoy you come to the town of l'Aiguillon. On the right bank of the river, La Faute-sur-Mer, there is a midget harbour among the reeds. The name on the port *balise* at the entrance is 'Gounelle de Virly'. On the opposite side of the river is what the *Syndicat d'Initiative* refer to as their *port de plaisance*. If you can get the lock gate to open you have a small pond which I am assured never has less than 1·8m. Makes a change from all that open sea, doesn't it? And when you are knee deep in mussels and oysters, well . . .

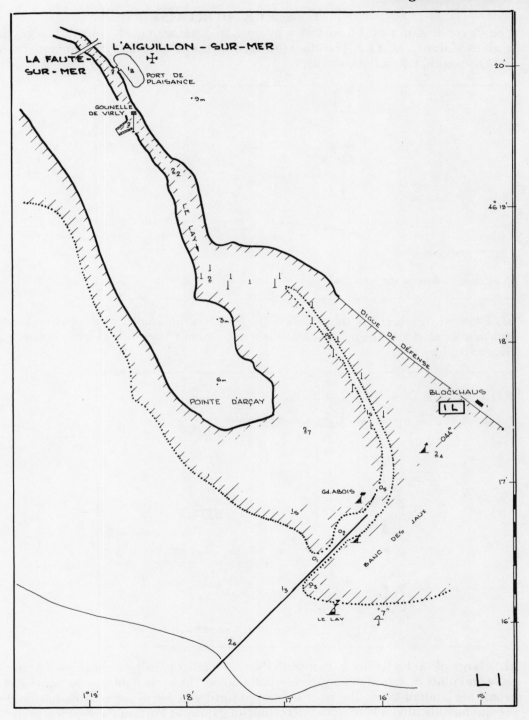

LA SÈVRE-NIORTAISE

You've got mud in your bloodstream by now, so come with me among the sedges and wild birds to Marans, chart L2. The place to start off is the lit landfall buoy, but first, if coming up from the south, take a back mark,

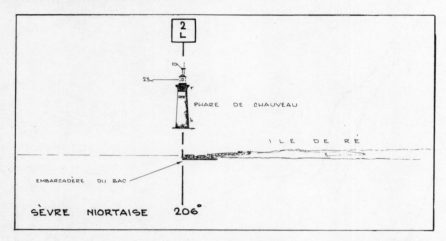

2L Cheveau lighthouse × the middle of the ferry terminus on Ile de Ré. In my sketch I have shown it for clarity to the left instead. There is another back mark which takes over near the river entrance,

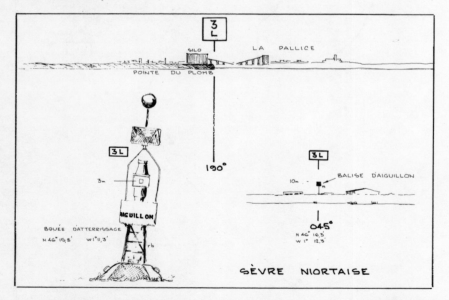

3L a large silo at La Pallice × Pointe du Plomb. Further to help establish your position is the *balise* on Pointe d'Aiguillon and the landfall buoy of the same name on the same sketch, 3L. From this landfall buoy, the river is buoyed until you get to the narrow part and should present no difficulty. There is an interesting long quay at Port du Pavé—view of its slip on

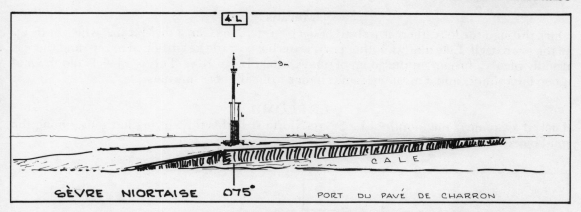

4L 3km away from the town of Charron. About here the channel buoys start, running all the way to the lock at Brault. The stb'd numbers finish at '31' and the port at '42'. The lock opens 2 hours before to 2 hours after local high water at springs (1 hour before to 1 hour after at neaps); the tide here is 20 minutes later than at La Rochelle. The lock, though large, has unpleasantly sloping sides which could be dangerous in any wind. One km before the lock is a remotely operated bridge. The starting button is a matter for the eagle eye of the *éclusier* at Brault. Upstream of Port du Pavé the river dries *1·0m,* soft mud.

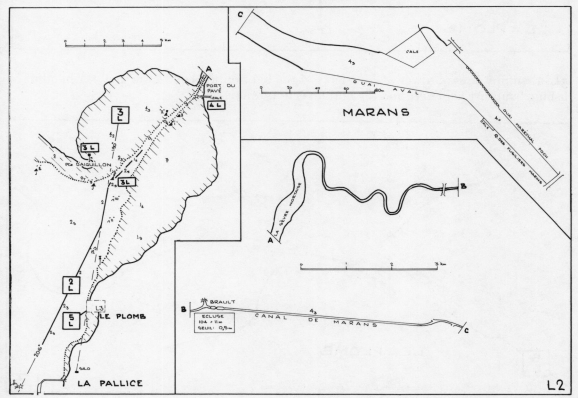

105

Marans

Once through the lock, there is a straight deep stretch, the Canal de Marans, which ends up in the town itself. Like many another port, waterborne trade has dropped to zero and there is usually plenty of room on the 500m of quays, chart L2 (p.105). This is a lovely old town of 4,000 inhabitants and a most welcome change from all those marinas.

LE PLOMB

Tucked away near the mouth of La Sèvre-Niortaise is this tiny drying harbour among the mud banks, chart L3. One mark takes you there

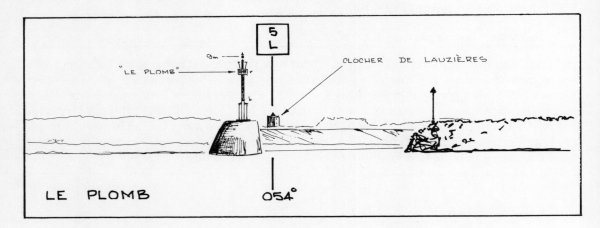

5L a stumpy square tower at Lauzières × the light on the end of the quay. A handful of fishing boats are kept here and the bottom is thick mud.

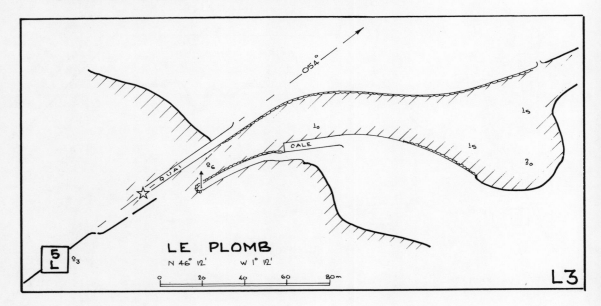

ILE DE RÉ
St Martin-de-Ré

From 1600 to the Napoleonic era, Ile de Ré was the scene of ding-dong rivalry between the French and the English. I don't know who won, but who cares with a lovely little harbour like St Martin ready to welcome you, chart L4?

The main entrance starts at the safe water lit buoy 'Rocha' and is

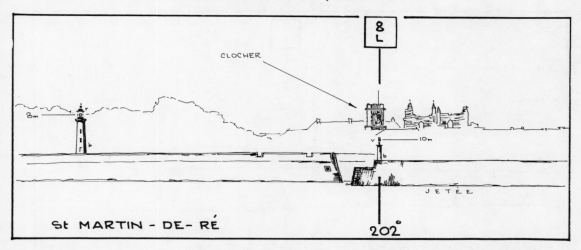

St MARTIN - DE - RÉ

8L St Martin church ✕ the light at the end of the west jetty. A cable or so outside the harbour, several mooring buoys have been put for use while waiting for the tide. The outer harbour is for fishing boats except for the north end of the Quai de Bernonville. If you wish to lock in to the inner basin, which is often very crowded, the gates open at a tidal rise of between 3·0 and 3·2m, daytime only. Ile de Ré is served by the short ferry between La Pallice and Sablonceaux

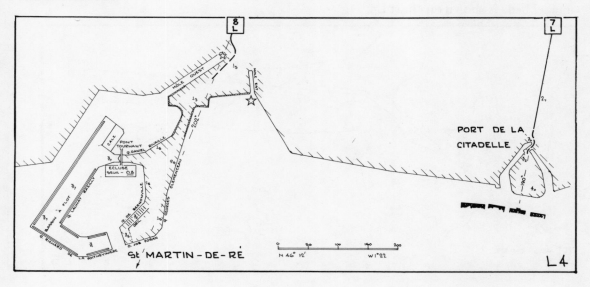

St MARTIN - DE - RÉ

L4

Port de la Citadelle

500m east of the harbour entrance there is another little harbour, not a joke one this time. It was built 100 years or more ago for the convenience of convicts from the nearby prison who took the package deal voyage to Devil's Island, Guinea. The harbour is on chart L4, and if you don't mind looking at the guards of the prison through the wrong end of their binoculars, you'll have the place to yourself. The mark is

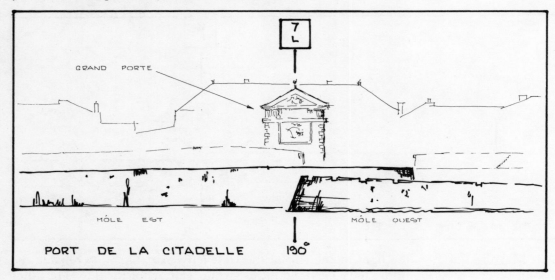

7L the ornate doorway × the west mole.

Loix

This is a tiny but ancient little harbour at the head of the Fosse de Loix. A way through the mussel beds is shown on chart L5.

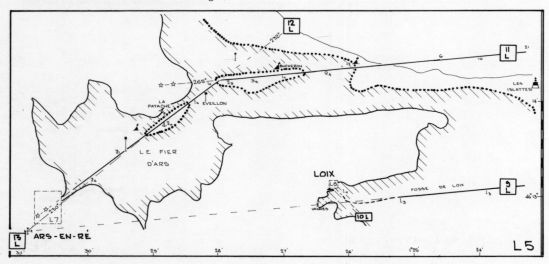

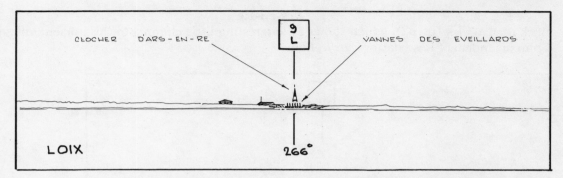

9L Ars church × a group of locks at Eveillards. This mark brings you to where the withies take over which lead you on to chart L6, the final mark being

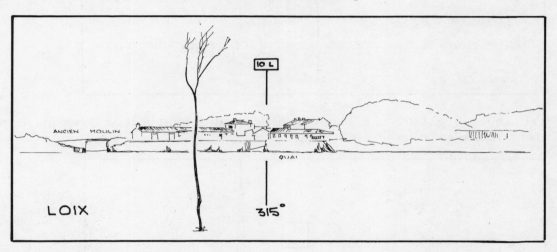

10L the top of the slip at 315°.

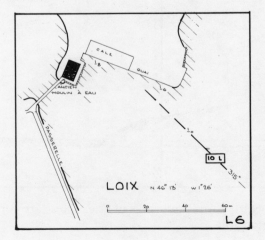

French Pilot 4

Ars-en-Ré

Back to chart L5 (p.108), which shows an interesting little channel to this ancient village. From just north of Les Islattes *tourelle*, find

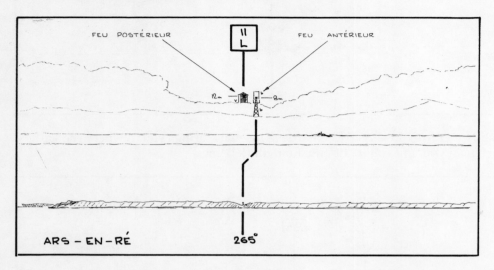

11L a hut × a light structure. This line, offset for clarity, must be accurately kept until the channel marks come up,

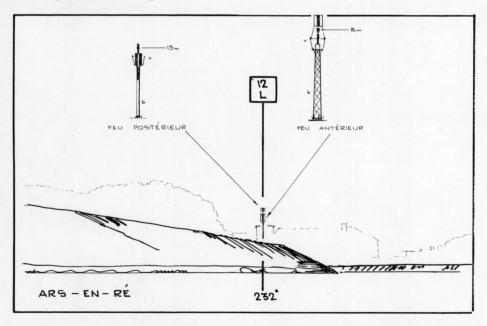

12L two light structures with panel daymarks in line. These are very difficult to find as they are 2½ miles away so I give you

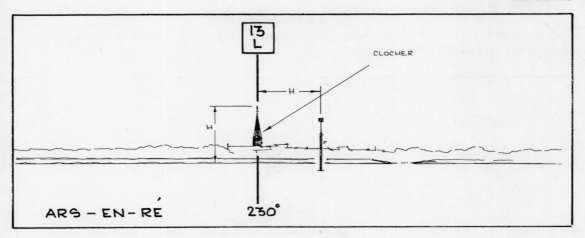

13L Ars church to the left of the half-way port *balise*. This is almost as good for the first half of the correct transit. There is a popular deep water anchorage at La Patache opposite what used to be an old ferry landing. The last quarter mile is canalised, shown on chart L7. The lock gates don't seem to work any longer, remaining permanently open, but there are several useful quays as you will see.

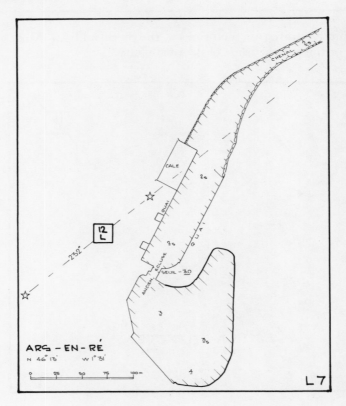

French Pilot 4

La Flotte

This friendly little fishing village welcomes visiting yachts (chart L8). The one mark is the same as the sectored white light.

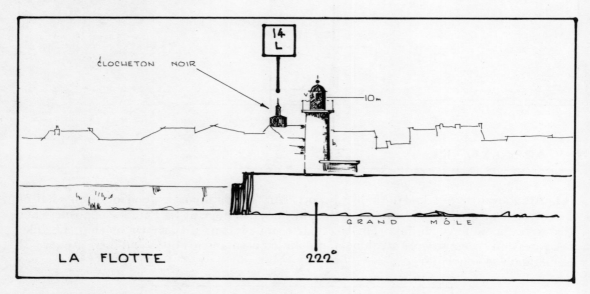

14L A black distinctive spire (not a church) ✕ the pierhead light. Most visiting yachts use the outer harbour or the seaward side of the north jetty.

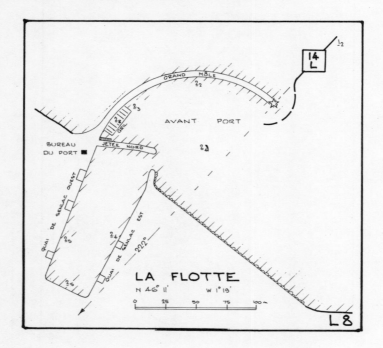

Rivedoux

Small, shallow and muddy could well describe this harbour (chart L9), but it's a handy waiting place before going into La Rochelle. The mark is

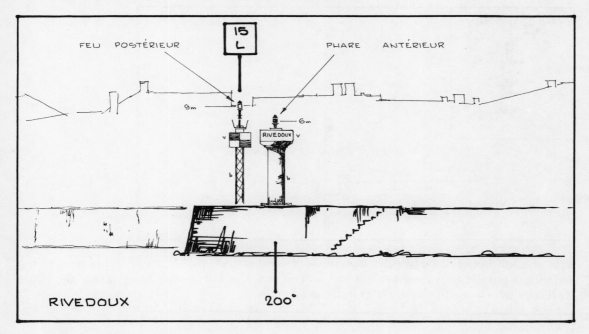

15L the two harbour lights in line.

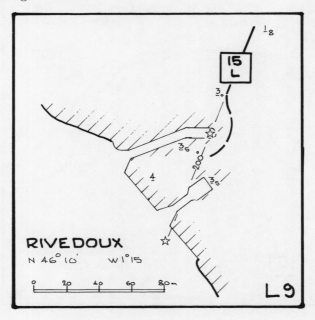

LA PALLICE

This enormous harbour is hardly the place for pleasure yachts but this book would be incomplete if I didn't show it on chart L10. There is one big-ship mark

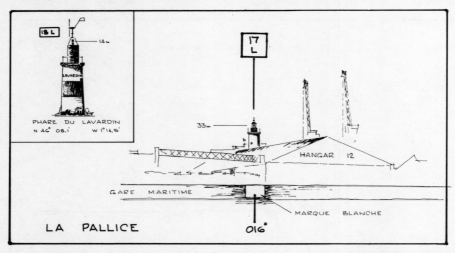

17L the lighthouse × a white mark on the south end of the Gare Maritime. If you are driving a 100ft gin palace, they might relent and let you in to the wet basin, but beware of the Ile de Ré ferries which use the harbour as if on speed trials.

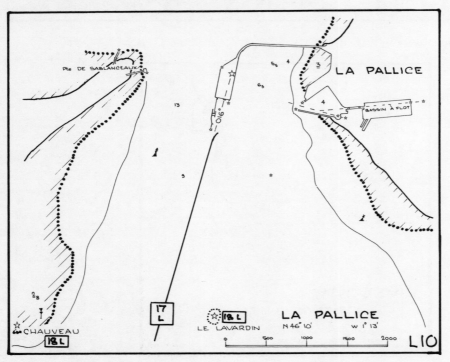

LA ROCHELLE

Steeped in history and flattened three times since 1500, this town took its 1944 bombard-ment as a matter of course. The entrance, between two towers, must be one of the most well-known, chart L11.

Portraits of four of the most important lighthouses in the Pertuis de l'Antioche are on

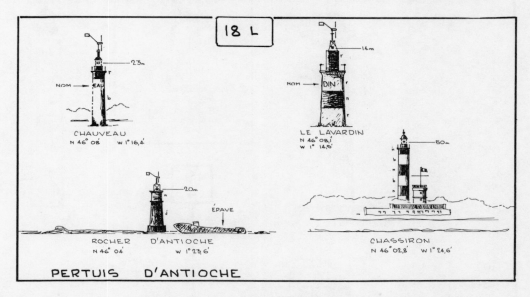

18L Chauveau, Le Lavardin, Rocher d'Antioche and Chassiron. Our main mark is

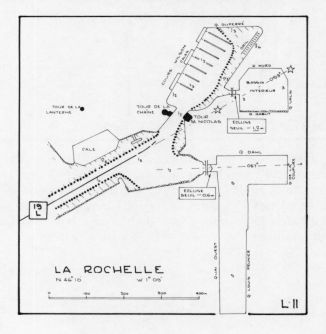

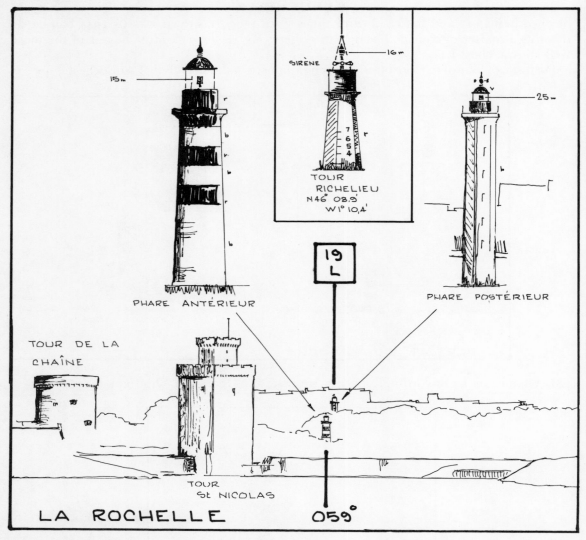

SIRÈNE — 16 m

TOUR
RICHELIEU
N 46° 08.9'
W 1° 10.4'

15 m — PHARE ANTÉRIEUR

25 m — PHARE POSTÉRIEUR

19
L

TOUR DE LA
CHAÎNE

TOUR
St NICOLAS

LA ROCHELLE   059°

19L the two lighthouses (lit all day) in line. This transit leaves Richelieu close to port, pictured on the same sketch. From here onwards, the channel is dredged to 0·2m though it deepens when you get on to chart L11. Visiting yachts use the pontoons by Cours Wilson, near the *1·2m* sounding or may lock into the Bassin Intérieur. The gates are open by day 2 hours before to ¾ hour after high water and at night 1 hour before to ¾ hour after.

La Rochelle has the facilities of a seaport of 80,000 inhabitants.

PORT DES MINIMES

Though a long way from the town, this marina, directly opposite the Tour Richelieu, is most convenient. It has all the usual *port de plaisance* amenities, chart L12.

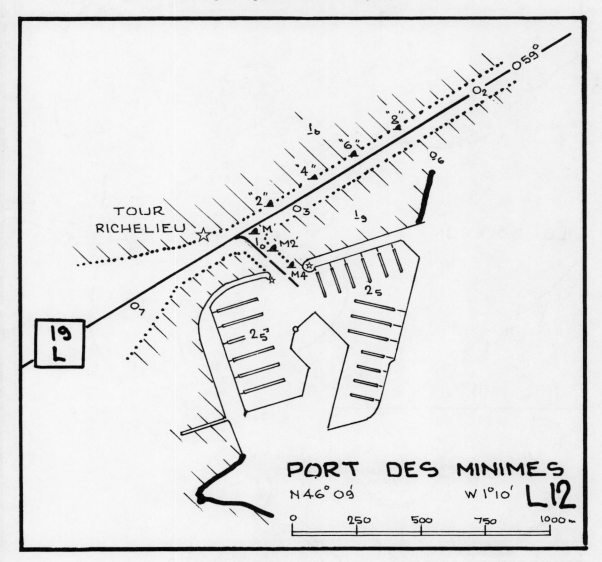

PORT DES MINIMES

N 46° 09'   W 1° 10'   **L12**

We are now going out into the Pertuis d'Antioche, chart L13.

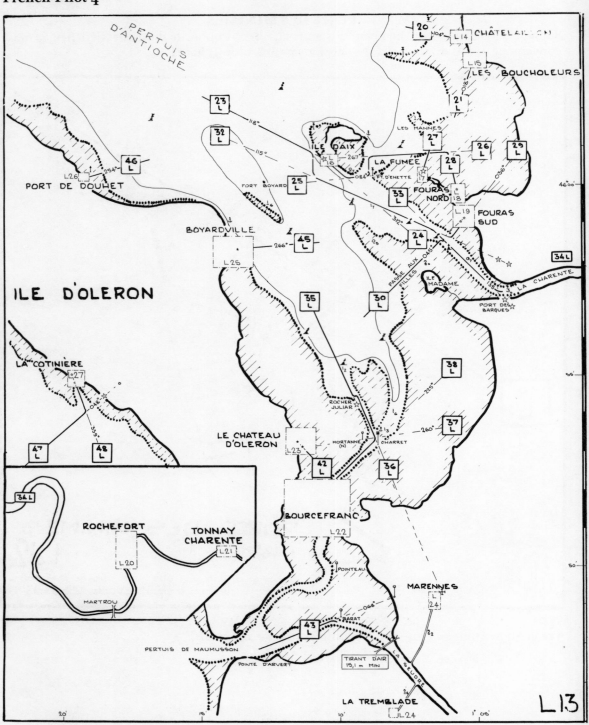

ILE D'OLERON

CHÂTELAILLON
This and the next are a couple of midget drying harbours protected by recently constructed rubble breakwaters, chart L14. A single entrance mark suffices

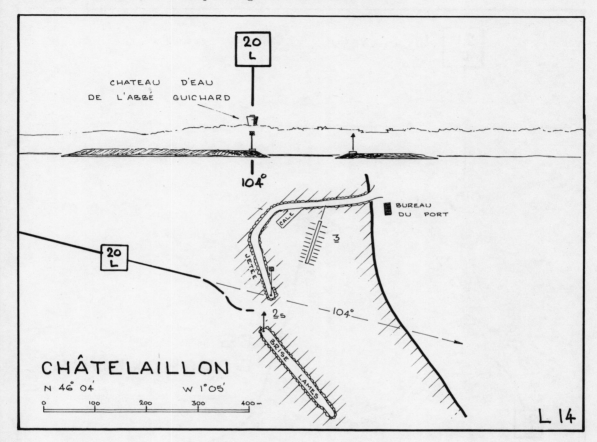

20L a water tower ✕ the port hand *balise*.

LES BOUCHOLEURS

As you will see from chart L15, the original quay has been extended and a single mark is enough,

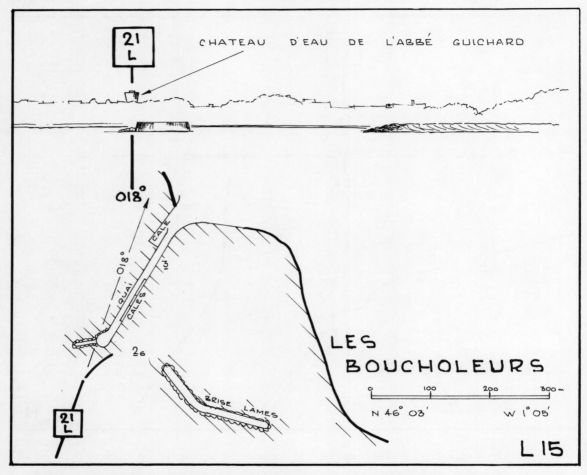

21L the same water tower ✕ the left hand of the end of the quay. The slip is used by the oyster men's tractors but the east side of the quay is clear.

ILE D'AIX

Napoleon passed his last hours on French soil on this island. Its best approach, chart L13, (p.118) is on line

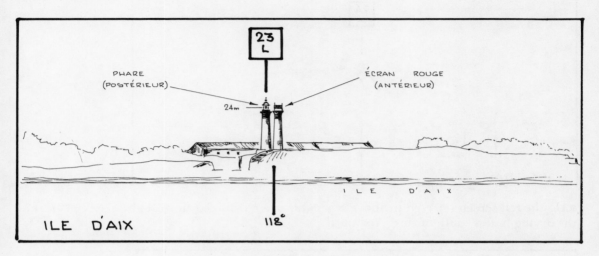

23L two lighthouses open as shown. If you look at chart L16 you will see that the nearest light is no lighthouse but a support for a red glass screen—a novel, if somewhat expensive, way of providing a sectored light. When ½ mile off this bizarre pair, come south-about of the island. There are many mooring buoys in this area. The Jetée des Barbotins is reserved for the ferries from La Fumée and elsewhere. The whole of the SE of the island dries.

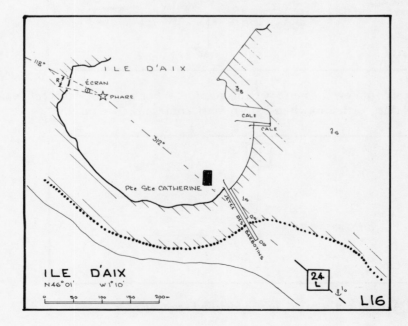

French Pilot 4

When leaving the anchorage near the jetty on Ile d'Aix to go round to the next harbour, La Fumée, use

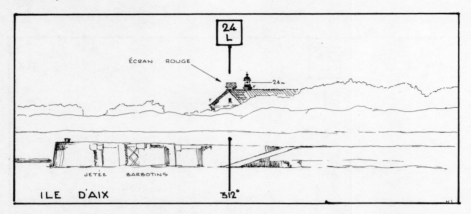

24L the red screen referred to above × a gable of a red-roofed store. This takes you clear of the drying banks south of Ile d'Aix, until

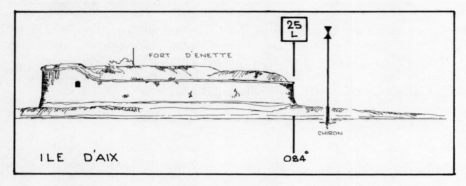

25L the right side of Fort d'Enette × Chiron *balise*. When a cable off this *balise*, take the deep water channel due north and when a stern mark comes up, turn on

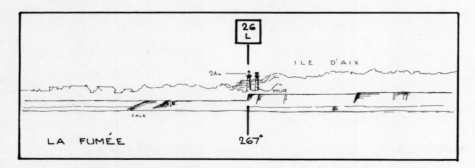

26L the Ile d'Aix light × a break in the fortified sea wall.

LA FUMÉE

This is the ferry terminal for Ile d'Aix but one can dry out at the root of the quay, chart L17.
The mark is

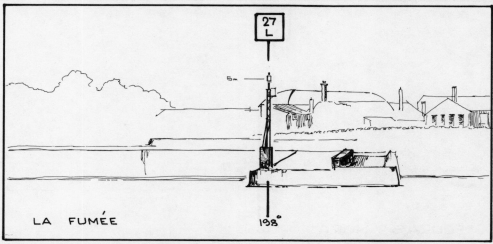

LA FUMÉE

198°

27L the left-hand end of the hemispherical roof ✕ the light at the end of the slip.

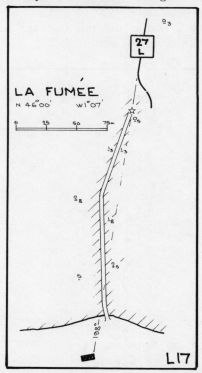

LA FUMÉE

N 46°00' W1°07'

0   25   50   75m

L17

Continuing eastward along the previous mark, see view 26L, one can branch southward to a
tiny harbour.

PORT NORD DE FOURAS

Chart L18 shows all the details. The mark is

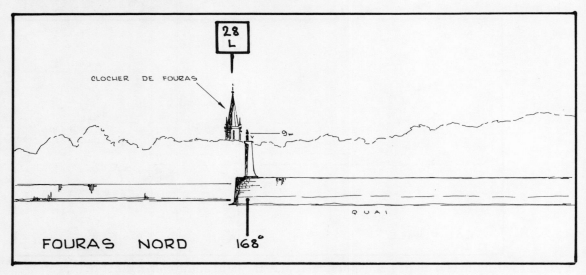

28L Fouras church ✕ the pierhead light.

We have now gone as far as possible into this bay, the Anse de Fouras, so how about retracing our steps to the south side of this peninsular?

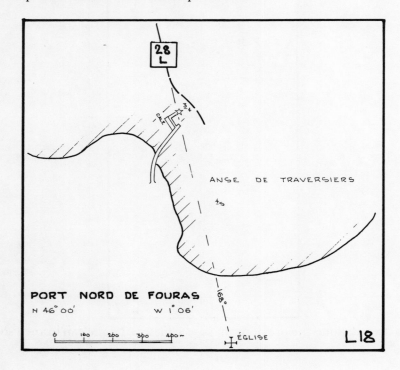

PORT SUD DE FOURAS

The best approach from the Pertuis d'Antioche is to use the river marks for the river Charente, chart L13 (p.118), but I'll give you the single entrance mark now, chart L19.

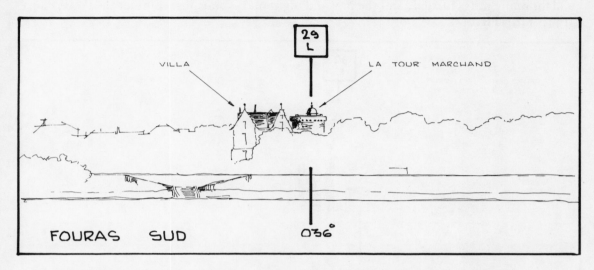

29L La Tour Marchand × a prominent villa near the beach. The harbour is mostly used by fishing boats, but there are several drying moorings in the bay.

The two leading lights on chart L19 are no direct concern of the Fouras entrance: they are

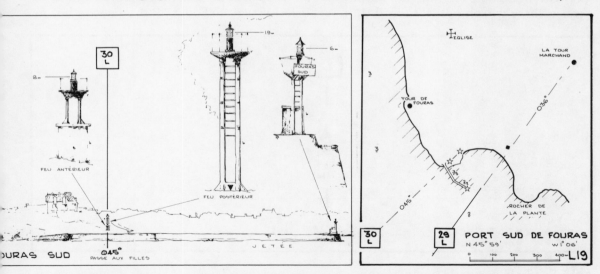

30L two light structures in line which are used for the Passe aux Filles on the south side of the river mouth of la Charente.

RIVER CHARENTE

Nearly 16 miles up the La Charente from Ile d'Aix is the lovely old fortified town of Rochefort, the one-time headquarters of the sea defences of this entire region. The river is tidal for many miles further, and the day and night marks start just south of Ile d'Aix, chart L13 (p.118). First take

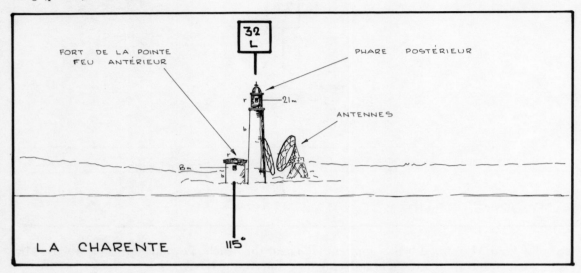

32L two lighthouses in line. Behind the enormous back light, somewhere in the middle of France, there are some great radio dishes. Just before you get to Fouras Sud, come to stb'd on

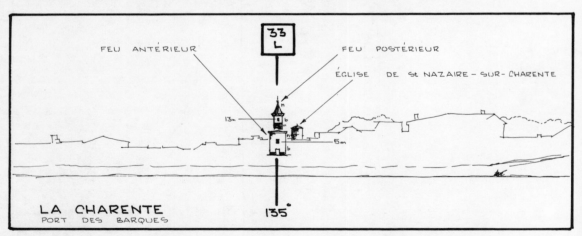

33L a pair of low but distinctive lights in Port des Barques. There is a slip and small quay at this rather gone-out-of-business riverside town. The rest of the river is well marked.

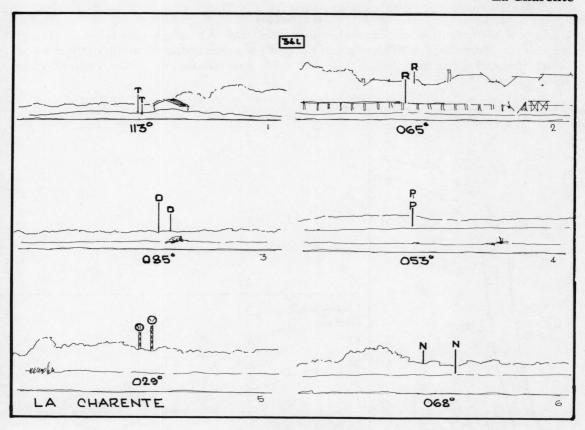

113°   065°   2

085°   3   053°   4

LA CHARENTE   5   068°   6

34L shows the daylight alignments for the first few bends. These are really for ships and it is sufficient to keep roughly in the middle of the river. The sequence runs 1 to 6, and so on.

The new lifting bridge at Martrou is 1½ miles downstream from Rochefort and replaces the adjacent disused transporter bridge which has a minimum headroom of 45m. The new bridge has a minimum headroom of 2m when down and 30m when up. From May to September inclusive, the bridge is raised at 05.15 and 20.45 every day, so be ready at those times. At other times it opens only for commercial traffic, in which case you can slip through at the same time. The signals for the bridge to be lifted are: when going upstream 3 long blasts; when coming downstream you must give 2 long blasts twice. The time taken to raise or lower the bridge fully is 2 minutes. On both upstream and downstream sides of the south tower of the bridge there is a panel of five signal lights, arranged in two groups. One group has three vertical white lights; the other only two, a red and a green.

Red and Green .... Stand by, bridge down or descending
Red and Green alternating .... stand by, bridge ascending
Three white fixed .... passage permitted, bridge up

Rochefort

The *port de plaisance* at Rochefort is shown on chart L20. The single gate opens, with luck, about 1 ½ hours either side of high water. There is a pontoon on the north side of the entrance which is useful if you have to wait. Rochefort is an interesting old walled town of 35,000 inhabitants.

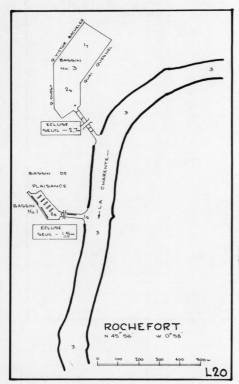

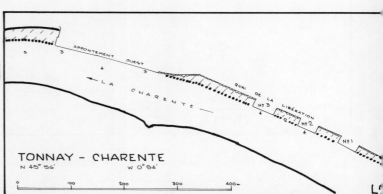

Tonnay-Charente

Three miles further up the river are the deserted quays of this once prosperous riverside town, chart L21. Now, grass grows on the railway tracks and the dock side cafés are deserted. All the wharves are wooden piles, very dangerous for small boats, but an excellent place to moor is fore and aft on the Quai de la Libération between nos. 2 and 3 platforms.

BOURCEFRANC

It's now time to take a look at the Coureau d'Oléron which extends from Ile d'Aix down to the new road bridge at Bourcefranc. The direction of the Lateral buoyage is with the flood, i.e. northward. Back then to chart L13 (p.118) and stand by for our first mark.

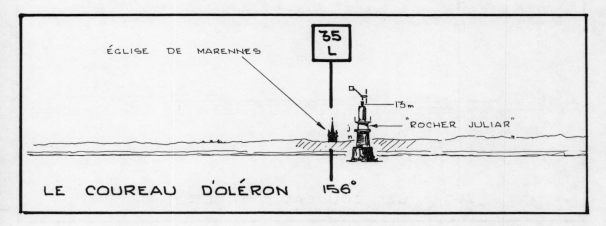

35L Marennes church × Rocher Juliar lighthouse. When 100m from the light, come a few metres eastward on to

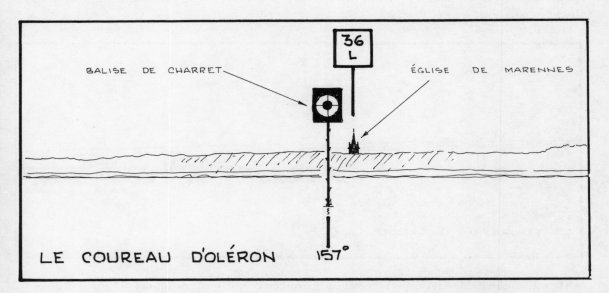

36L the same church of Marennes × Charret, a distinctive looking *balise*. This takes you past Rocher Juliar to the breast mark

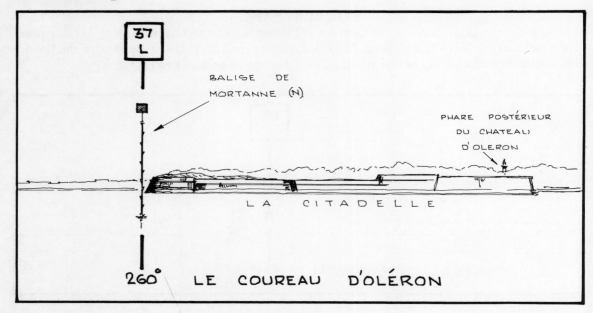

37L the left wall of the citadel of Château d'Oléron × the *balise* of Mortagne Nord. At this point is an excellent place to anchor but if you wish to carry on, look towards the new bridge for

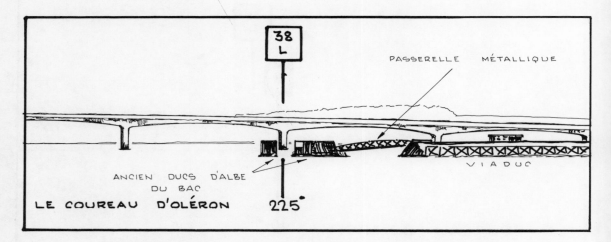

38L one of the bridge piers seen between the old ferry terminus dolphins. There is only one pier which fits anything like this bearing.

Transfer now to chart L22 and you will see this mark takes you just to the NW of the stb'd buoy named 'Craze'. Leave this to port and follow the track shown past La Corde. Here, just south of La Corde, is another excellent place to anchor. The harbour for Bourcefranc is called Le Chapus and a single mark is sufficient

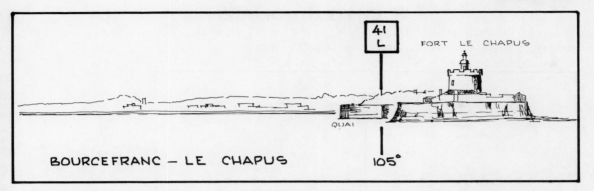

41L the harbour entrance seen to the left of Fort le Chapus. The harbour is rather cramped and used by only a few fishing boats.

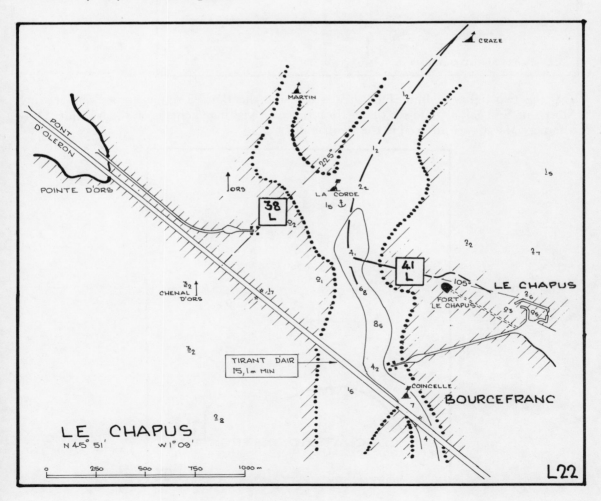

CHÂTEAU D'OLÉRON

This is an interesting little harbour built at the same time as the citadel, chart L23. Retrace your steps to the Craze buoy and come in on the transit

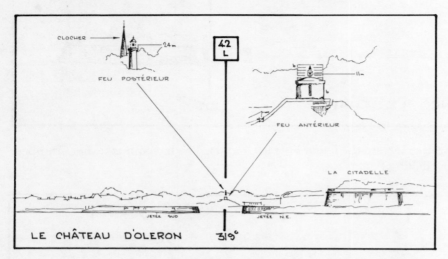

42L the two lights in line. This means crossing the Banc d'Agnas (*0·3m*) and leaving Mortagne Sud *balise* 5m to stb'd. The best place for visiting yachts is on the NE side. Only commercial craft are allowed in the wet basin.

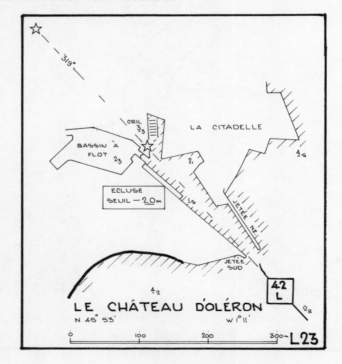

RIVER SEUDRE

From here southward it is an adventurous but rewarding trip up this river, finally to come either to Marennes or to La Tremblade. There are no marks that I can give you but the channel is shown on chart L13 (p.118) and is very well buoyed. You can start off on chart L22 (p.131) at the port buoy 'Coincelle' and carry on SE under the adjacent arch of the road bridge. This navigable arch is marked by lights in the middle at both sides of the bridge plus port and stb'd panels (red square and green triangle). Just north of the Pointe d'Arvert there is a mark for which special beacons have been erected.

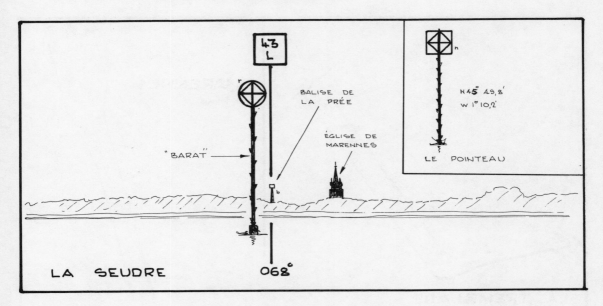

43L *Balise* de la Prée × Barat and its function is to take you almost into the river entrance which from then on is buoyed.

Here is as good a place as any to warn you off using the Pertuis de Maumusson. I have been through once in total ignorance, near the high water, and thought nothing of it; but every time I relate this I listen to tales of strandings, drowning, buoys being out of position, etc. If you think your nerves are up to scratch, just stand on the Pointe d'Arvert when the ebb is running against a Force 2 zephyr. The sight, if you are still set on going south, should dispel the boredom of the long haul round the north end of the Ile d'Oléron. However, back to the tranquil river Seudre.

French Pilot 4

Marennes

There is no difficulty getting up to the lock in Marennes, chart L24. This is available from half tide upwards. What was the old wet dock is now a *port de plaisance* and very tranquil too. What a pity it is so far from the sea.

La Tremblade

La Tremblade has no lock but an electric cable crosses the canal with a minimum height of 15·2m.

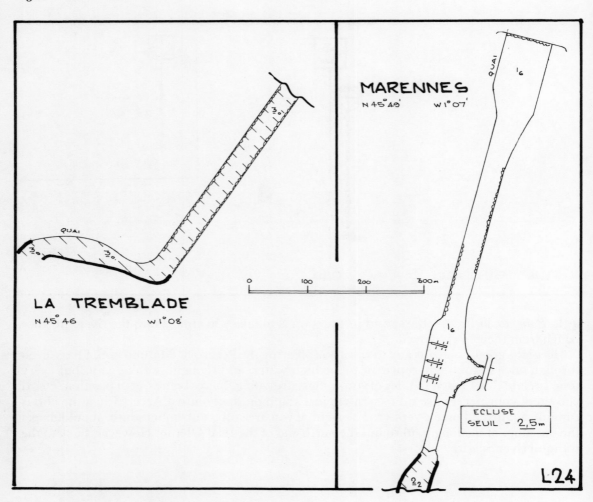

ILE D'OLÉRON
Boyardville

We now go back to the Pertuis d'Antioche to carry on westward round Ile d'Oléron. The old name for this recently refurbished harbour, chart L25, was La Perrotine and the entrance mark is

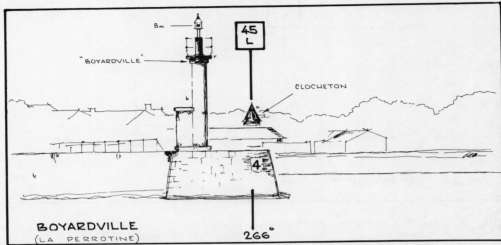

45L a spire just to the right of the pierhead light. As soon as the light is reached, carry on parallel to the breakwater about 10m off. The *port de plaisance* has a simple pair of automatic gates which open when the rise of tide is 2·4m above chart datum. During the movement of the gates, a red light shows on both sides. Don't forget that they close automatically too. One or two yachts have tried to play 'last across' and have been chopped in half.

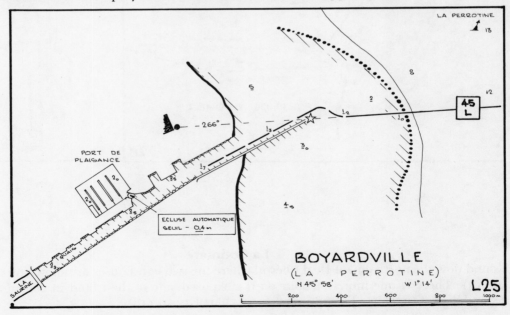

Port de Douhet

This is built on the mouth of a small stream and has recently been extended to form a useful little *port de plaisance*, chart L26. One mark avoids the rocks to the north

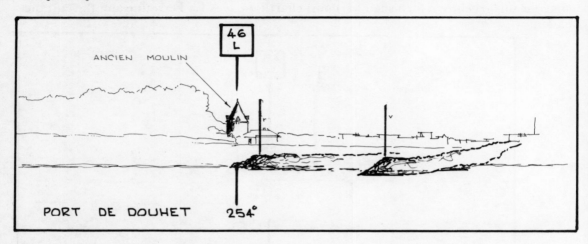

46L an old mill × the port hand entrance mark.

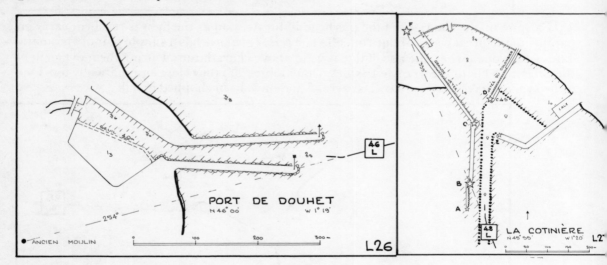

La Cotinière

Round now the north end of Ile d'Oléron where the two lighthouses are shown on view 18L (p.115). There is one more harbour on the exposed side of the island in constant use by a small fleet of trawlers. The approach is on chart L13 (p.118).

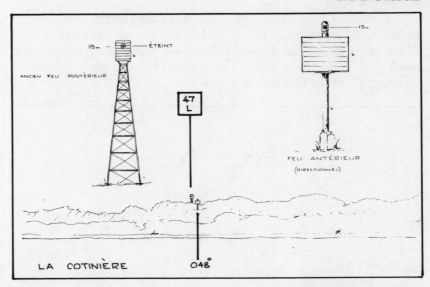

47L a pair of light structures in line. The back light has been extinguished but the tower remains for daytime use. The front light is now directional and has a large white panel for use during the day.

About ½ mile from the shore, look north for the final line

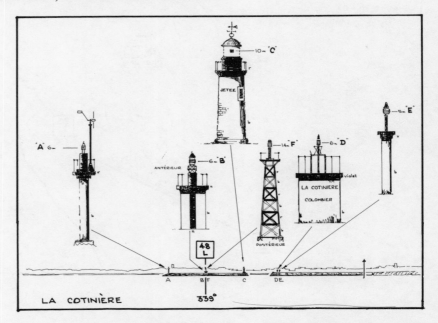

48L two lighthouses in line. The harbour is on chart L27 and, to avoid confusion, I have had to give the six lights letters of recognition. 'F' and 'B' are the ones we need for this mark. To get out of the way of the fishing boats, try the north end of the Grande Jetée.

LA GIRONDE

Before going up the river Gironde, I will show you the two main entrance marks, chart L28. While these are normally for big ship use, you don't have to wander very far off them in a small boat to be in trouble. The whole of the estuary is dominated by the Renaissance architecture of the Cordouan lighthouse.

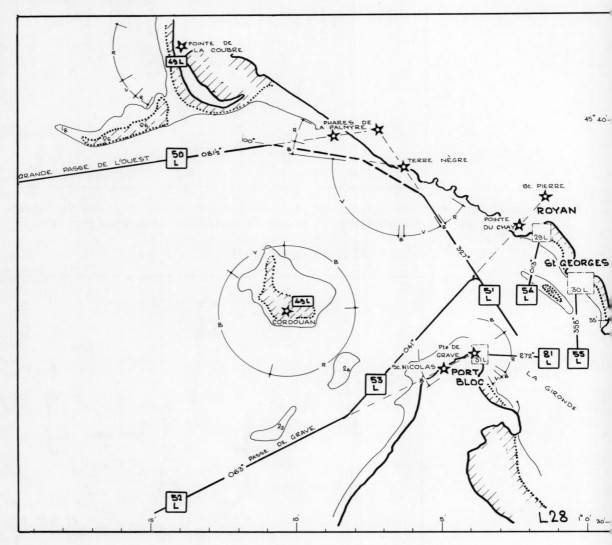

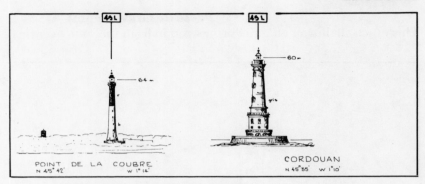

49L A portrait of Cordouan and Pointe de la Coubre lighthouses.

Grande Passe de l'Ouest (12·8m)

In recent years this has been shifted and redredged and starts out to sea at the landfall light float 'BXA',

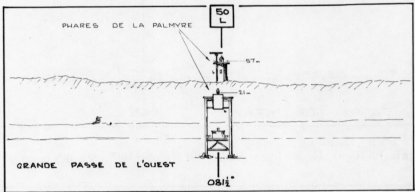

50L the two lighthouses of La Palmyre. The back one carries a radar scanner and the front one is knee deep in the river.

By night the sectors of the Terre Nègre light take you into the river and on to a back mark,

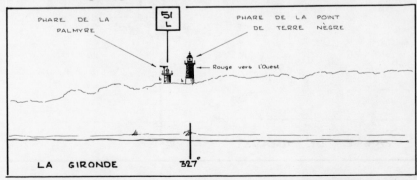

51L Palmyre light × Terre Nègre light. This takes you on to the main buoyed channel all the way to Bordeaux.

Passe de Grave (2·9m)

This much shallower channel brings you in from the sunny south,

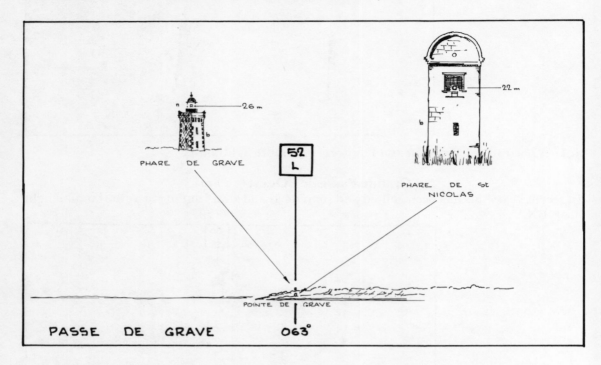

52L Pointe de Grave lighthouse × St Nicolas light. 3 miles from the front light take

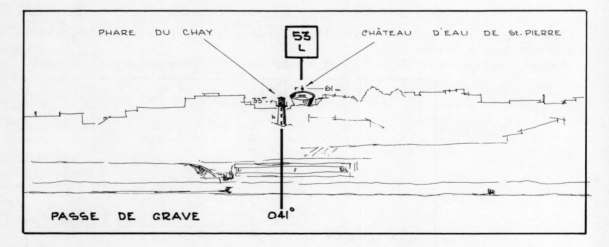

53L the water tower at Royan × the Pointe du Chay light. The back light is cleverly situated in the tulip-shaped top of the water tower. This mark soon joins view 51 L (p.139).

Royan

This is the usual halt for boats going in and out of the Gironde. The town has been completely rebuilt since the war and now has a commodious marina, chart L29. To avoid the Banc St Georges chart L28 (p.138).

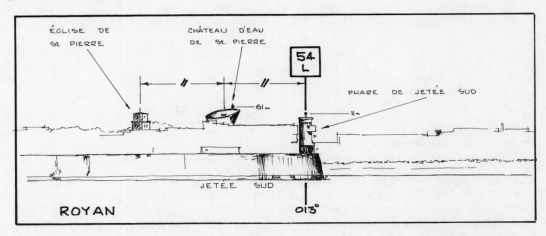

54L the south jetty light to the right of the church and water tower in the proportions shown. This is the mark the Port Bloc ferries use. Unless you choose to dry out to the east of the harbour, it is better to go straight in to the *port de plaisance*, but take note first of the drying entrance (*0·5m*), the undoing of many strangers.

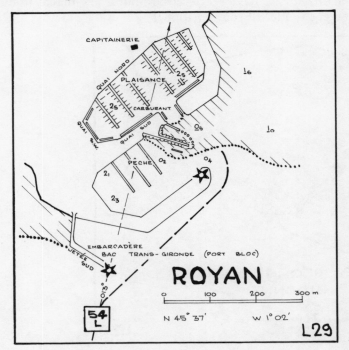

St Georges-de-Didonne

A muddy but otherwise convenient fishing harbour if you don't like marinas, chart L30. To avoid the south end of the Banc de St Georges, use

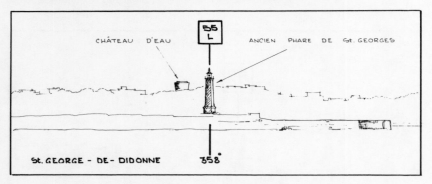

CHÂTEAU D'EAU — 55 L — ANCIEN PHARE DE St. GEORGES

St. GEORGE - DE - DIDONNE — 358°

55L the water tower × the old lighthouse.

From here onwards up the river there are a succession of small harbours mostly relics of the days of waterborne wine exports. The order of all these ports is that we shall go up the river Gironde dealing with the right bank, up the Dordogne by the right bank to Libourne, down the left bank and up the Garonne to Bordeaux. We then come down the left bank of the Gironde, finishing up with Port Bloc. The whole of this is shown on chart L32 which also indicates the distances in km, measured from Bordeaux or Libourne, of the various points. These distances are shown on the French chart of the river. Since the river is wide and the banks somewhat featureless, I have scattered around a few places useful for recognition, from which you can get an idea of your position.

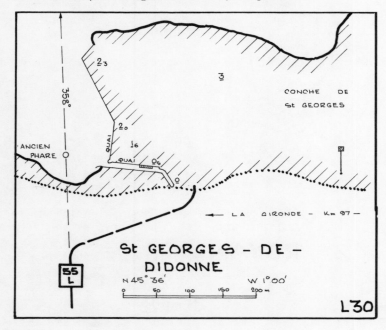

CONCHE DE St GEORGES

ANCIEN PHARE

QUAI

← LA GIRONDE - Km 97 →

St GEORGES - DE - DIDONNE

N 45° 36'   W 1° 00'

0   50   100   150   200 m

L30

La Gironde

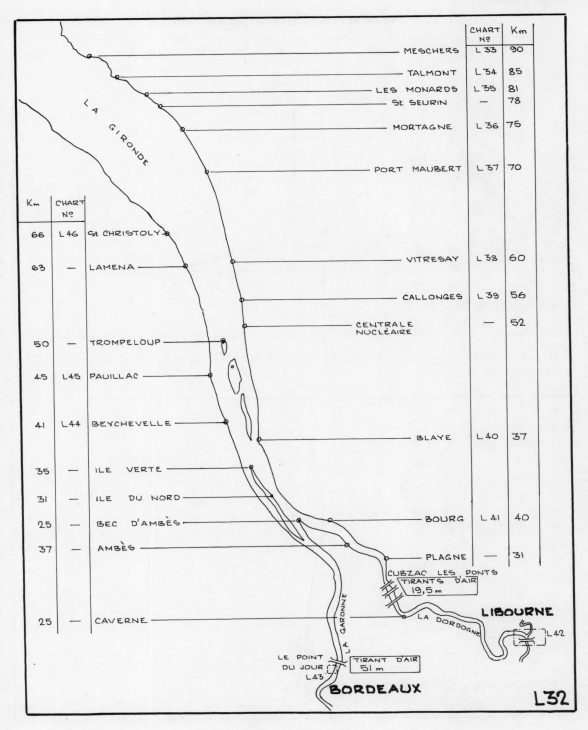

	CHART Nº	Km
MESCHERS	L 33	90
TALMONT	L 34	85
LES MONARDS	L 35	81
St SEURIN	—	78
MORTAGNE	L 36	75
PORT MAUBERT	L 37	70
VITRESAY	L 38	60
CALLONGES	L 39	56
CENTRALE NUCLÉAIRE	—	52
BLAYE	L 40	37
BOURG	L 41	40
PLAGNE	—	31

Km	CHART Nº	
66	L 46	St CHRISTOLY
63	—	LAMENA
50	—	TROMPELOUP
45	L 45	PAUILLAC
41	L 44	BEYCHEVELLE
35	—	ILE VERTE
31	—	ILE DU NORD
25	—	BEC D'AMBÈS
37	—	AMBÈS
25	—	CAVERNE

LA GIRONDE

CUBZAC LES PONTS
TIRANTS D'AIR 19,5 m

LIBOURNE
LA DORDOGNE
L 42

LE POINT DU JOUR
L 43

TIRANT D'AIR 51 m

LA GARONNE

BORDEAUX

L32

143

Meschers-sur-Gironde (Chart L33)

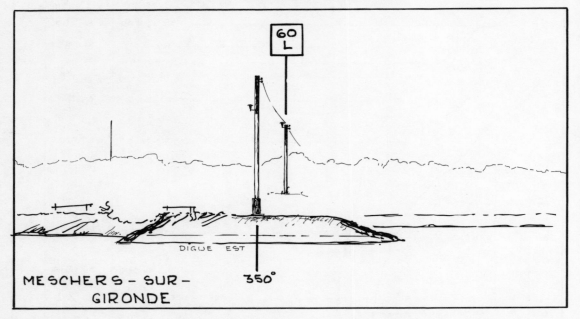

60L Two posts carrying lights in line. Disregard the stb'd *balise* 2 cables to the east as this serves for the main river.

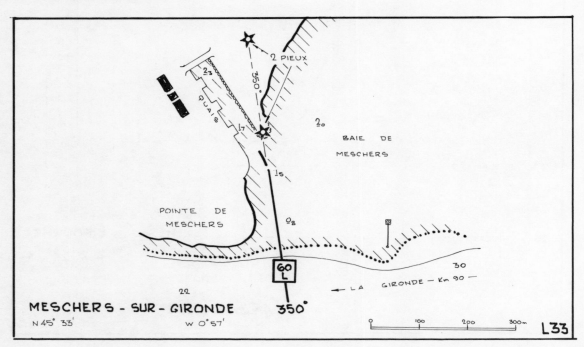

Talmont-sur-Gironde (Chart L34)

An anchorage only, in a convenient stopping place out of the tide. There are two or three mooring buoys laid but unfortunately in a very fierce part of the current. To get further in use

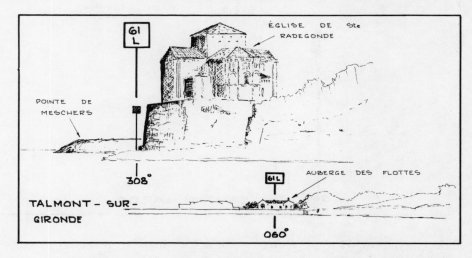

61L a port *balise* × the wall of the church, Ste Radegonde, which dates from the fifteenth century.

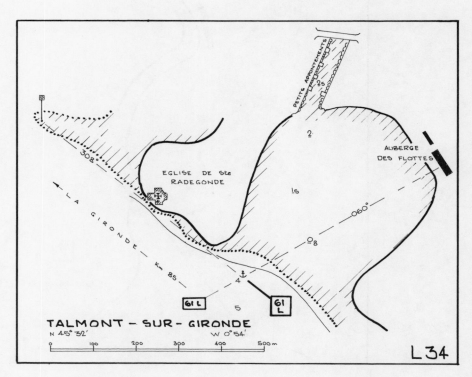

Les Monards (Chart L35)

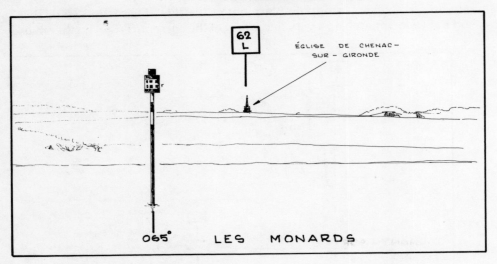

62L Chenac-sur-Gironde church × the *balise* which has reflective panels.

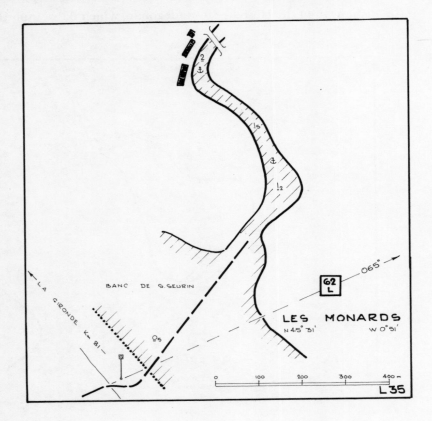

St Seurin d'Uzet

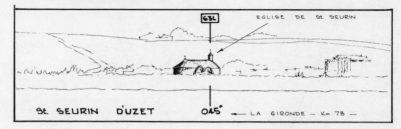

63L View only of the church.

Mortagne-sur-Gironde (Chart L36)

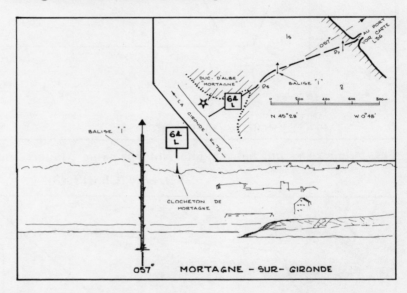

64L The church × *balise* '1'. The gate which only operates in daytime can be opened about 1½ hours either side of high water. There is a convenient drying quay just outside.

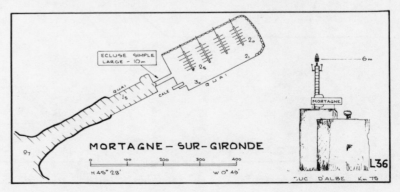

Port Maubert (Chart L37)

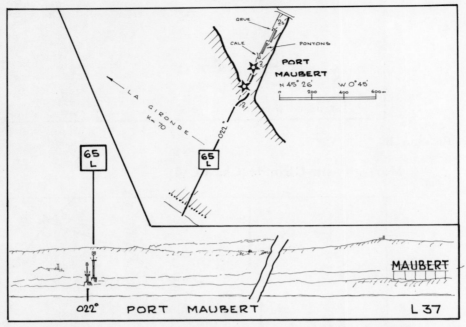

65L A pair of leading lights in line. Note the name 'Maubert' on the river bank.

Vitrezay (Chart L38)

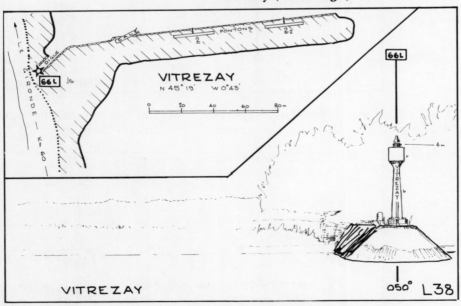

66L No marks are necessary, the view of the light is enough.

Callonges (Chart L39)

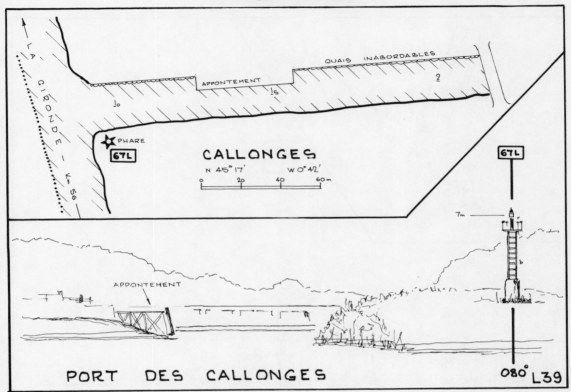

67L View only of entrance light.

Centrale Nucléaire

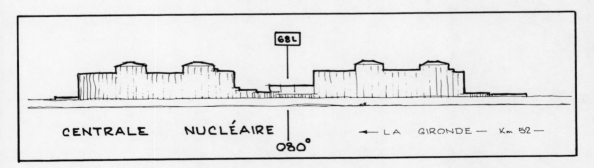

68L View only of the nuclear power station. There are many water intakes, etc., as much as half a mile into the river.

Blaye (Chart L40)

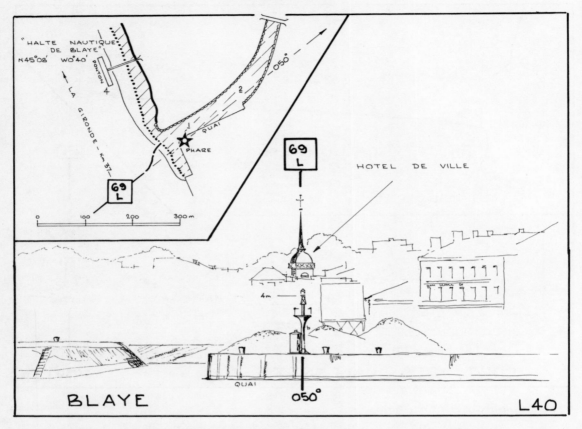

69L The Hôtel de Ville × the harbour light. There is a convenient alongside berth on the quay just east of the light, sometimes used by *sabliers*. A Ro-Ro ferry crosses the river from Blaye to Lamarque.

LA DORDOGNE RIVER
Bourg (Chart L41)

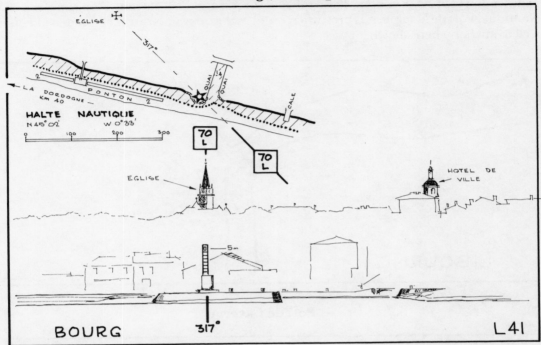

70L The church and the light in line.

Plagne

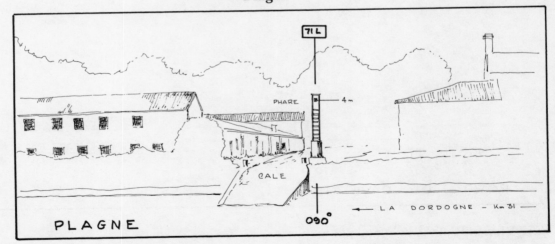

71L View of the light and slip.

The three bridges at Cubzac-les-Ponts offer no difficulty and the headroom is shown on chart L32 (p.143). The steel road bridge dates from 1882, the work of M. Eiffel.

Libourne (Chart L42)

This is a wonderful medieval town seldom visited by water. You are now in the middle of the wine country of St Émilion, Pomerol and Frontsac. The current alongside the Halte Nautique, Port de Nouquay, is pretty fierce and as the quay opposite is of wooden piles it is as well to anchor where shown.

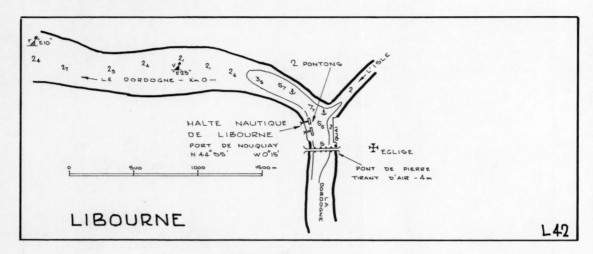

Port de Caverne

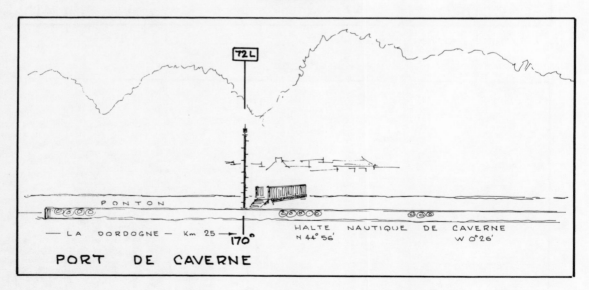

72L View only of the light and pontoon.

Ambès

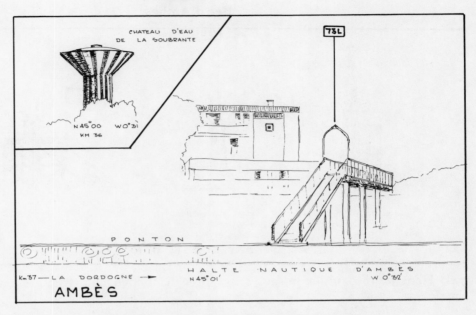

CHATEAU D'EAU
DE LA SOUBRANTE

73L

N 45°00   W 0°31
KM 36

P O N T O N

K.37 — LA DORDOGNE →

H A L T E   N A U T I Q U E   D'A M B È S
N 45°01'   W 0°32'

AMBÈS

73L View only of the Halte Nautique and the nearby remarkable water tower.

LA GARONNE RIVER

Le Point du Jour, Bordeaux (Chart L43)

Unless you are going through to the Mediterranean, there is nothing to attract small boats in the Bordeaux dock system, but the Halte Nautique is a convenient stop. There is a do-it-yourself crane for lifting masts in preparation for the Canal du Midi. The canal limits are: draught 1·6m, headroom 3m.

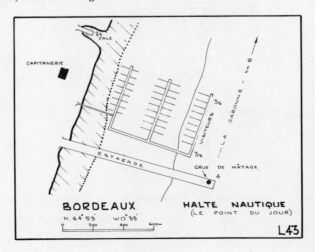

CALE

CAPITAINERIE

VISITEURS

LA GARONNE — KM 6

3s

3s

ESTACADE

GRUE DE MÂTAGE
4

BORDEAUX
N 44°53'   W 0°33'
0   200   400   600m

HALTE NAUTIQUE
(LE POINT DU JOUR)

L43

153

LA GIRONDE RIVER (Cont)
Bec d'Ambès

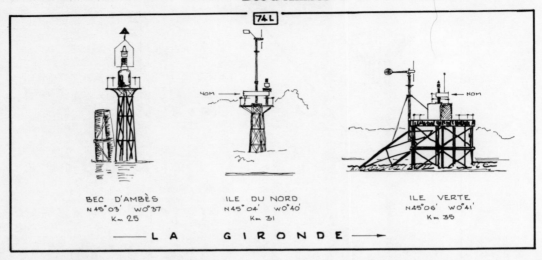

BEC D'AMBÈS
N 45° 03' W0° 37'
Km 25

ILE DU NORD
N 45° 04' W0° 40'
Km 31

ILE VERTE
N 45° 06' W0° 41'
Km 35

— L A G I R O N D E ⟶

74L Here, for recognition purposes, are three distinctive light structures on the way down the river.

Beychevelle (Chart L44)

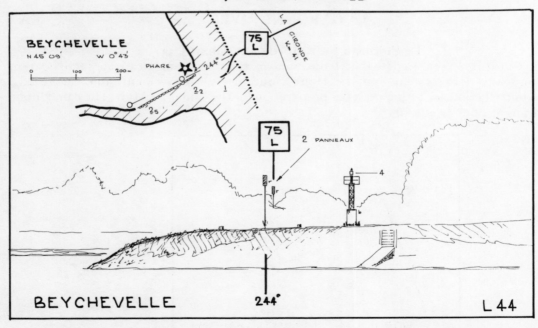

BEYCHEVELLE
N 45° 09' W 0° 43'

BEYCHEVELLE   L 44

75L Two red panels in line. What better place than this creek, so convenient for visiting the *chais* of the Château de Beychevelle.

Pauillac (Chart L45)

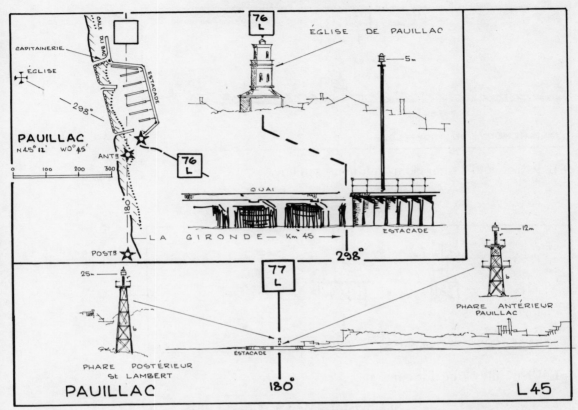

76L The church × the entrance light. Take care just around the entrance, both outside and inside, because the *estacade* is not continuous and allows the river current to circulate.

77L On the same sketch I show a pair of river leading lights, St Lambert × Pauillac. The *port de plaisance* has about 300 berths and room for visitors.

Trompeloup

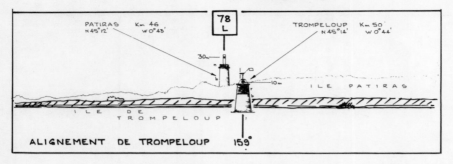

78L Patiras and Trompeloup lighthouses in line.

Lamena

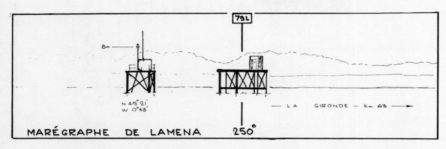

79L View only of the tide gauges.

St Christoly-Médoc (Chart L46)

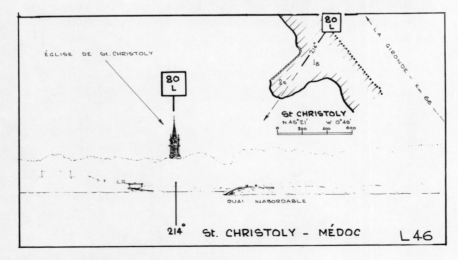

80L The church × the end of the quay.

Port Bloc

Back now on chart L28 (p.138) for this rather crowded little harbour whose dual function is to house the Ponts et Chaussées fleet, and to serve as the Royan ferry terminal. You don't really need a mark but there is least tide on

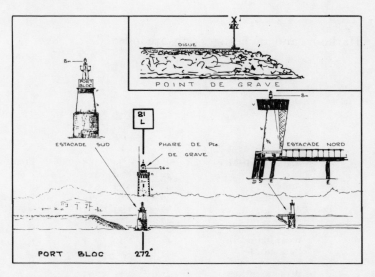

81L the Pointe de Grave light ✕ the south pierhead light. The harbour is on chart L31 and the only difficulty at the entrance is to keep clear of the ferries. If ever you have driven the wrong way down a motorway you'll know what I mean. The east quay is reserved for the Ponts et Chaussées, the ferries sleep at the *Ducs d'Albe*, and the pontoon belongs to the local yacht club, so your only hope is to find a place to anchor in mid-harbour.

If there is anything to be learnt in retrospect from a look at the over 100 harbours in this book it is the change in the maritime trade. Container ships berth in the few large harbours where formerly barrels were winched ashore and dozens of once prosperous inshore fishing harbours are now *ports de plaisance*. We shall never again see wine shipped from all those now sleeping ports of Charente.

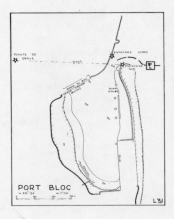

Index

Harbours and anchorages only are indexed.